AF540587

Practical Approaches in Horticulture

About the Authors

Dr. Rajaneesh Singh is presently working as Associate Professor & Head, Department of Horticulture, Tilak Dhari Post Graduate College, Jaunpur (U.P.). He did his Graduation in Agriculture from Udai Pratap Autonomous College, Varanasi, Post Graduation in Horticulture from GBPUA&T, Pantnagar and Ph.D. in Horticulture from BHU, Varanasi.

Dr. Singh has served Tilak Dhari Post Graduate College in various capacities like In-charge Garden Section, Captain in NCC, Member of the Sports Committee, Academic Counselor of IGNOU study center and Former Proctor of the college. Dr. Singh has 19 years teaching experience of horticulture. Dr. Singh has more than 62 publications to his credit (research papers 25, proceeding papers 5, conference/seminar papers 10, books 2, book chapters 10 and popular/semi-scientific articles 10). He had been chairman of the organizing committee of One National Workshop. Dr. Singh qualified ICAR NET in Horticulture and Vegetable Science and received Senior Research Fellowship in ICFRE during Ph.D. programme. He is fellow of Indian Society of Vegetable Science and Indian Society of Horticulture Science. He has supervised 4 M.Sc. (Ag.) and 3 Ph.D. students as Chairman of Advisory Committee.

Dr. Bijendra Kumar Singh is presently working as Assistant Professor, Department of Horticulture, Tilak Dhari Post Graduate College, Jaunpur (U.P.). He did his Graduation in Agriculture from VBSPU, Jaunpur, Post Graduation in Horticulture from BBAU, Lucknow and Ph.D. in Horticulture from BHU, Varanasi. He has specialization in Pomology.

Dr. Singh has served Tilak Dhari Post Graduate College in various capacities like Member of Campus Greenery of college under IQAC committee and Academic Counselor of IGNOU study center. Dr. Singh has 143 publications (research papers 30, proceeding papers 3, conference/seminar papers 50, book chapters 25 and popular/semi-scientific articles 35). He had been convener of the organizing committee of One National Workshop. Dr. Singh qualified ICAR NET in Fruit Science and received University Grant Commission (UGC) fellowship during his Ph.D. He is a life member of renowned societies and journals in India. Asian PGPR Society of Sustainable Agriculture, International College of Nutrition's, Society for Advancement of Research on Pomegranate, Trends in Biotechnology & Biological Sciences, Advances in Life Sciences and Trends in Biosciences. He has supervised 4 M.Sc. (Ag.) and 3 Ph.D. students as Chairman and Member of Advisory Committee.

Practical Approaches in Horticulture

Rajaneesh Singh
Head
Department of Horticulture
Tilak Dhari Post Graduate College
Jaunpur- 222 002
Uttar Pradesh

Bijendra Kumar Singh
Assistant Professor
Department of Horticulture
Tilak Dhari Post Graduate College
Jaunpur- 222 002
Uttar Pradesh

NEW INDIA PUBLISHING AGENCY
New Delhi – 110 034

NEW INDIA PUBLISHING AGENCY
101, Vikas Surya Plaza, CU Block, LSC Market
Pitam Pura, New Delhi 110 034, India
Phone: + 91 (11)27 34 17 17 Fax: + 91(11) 27 34 16 16
Email: info@nipabooks.com
Web: www.nipabooks.com

Feedback at feedbacks@nipabooks.com

ISBN : 978-93-89547-32-0

Composed and Designed by NIPA

World Noni Research Foundation

64, Third Cross Street, Second Main Road, Gandhi Nagar, Adyar, Chennai – 600 020
Telefax: 044-2442 3601 Email: mail@worldnoni.org Website: www.worldnoni.org

Dr. Kirti Singh
FNASc, FNAAS, FNABS
CHAIRPERSON

Foreword

Horticulture in current situation forms an integral part of food, nutritional and economic securities. India is recognized all over the world for its golden revolution. The country is second largest producer of fruits and vegetable after China. The present book entitled "Practical Approaches in Horticulture" cover all practical aspects of fruits crops, vegetable crops, herbaceous and shrubbery border, species and condiments, plantation medicinal and aromatic crops besides propagation methods and preparation of post harvest products like jam, jelly, marmalade, squash, tomato sauce, ketchup and chutney, pickle etc.

I am confident that this book shall be extremely useful to students and teachers of horticulture in the colleges and universities. I complement and congratulate the authors for their hard and sincere work in bringing out this publication at appropriate to meet the challenges of scientific horticulture.

Date: 05-09-2019
Place: Jaunpur

(Kirti Singh)
Former Vice-Chancellor

Preface

Undergraduate and postgraduate students of horticulture do not have any prescribed books in related to practical and laboratory exercise. The author is therefore making an attempt to meet this need. I hope that our book will helpful in standardizing subject matter for teachers, students.

I wish to express my sincere gratitude to Prof. Anil K. Singh (BHU), all faculty members of agriculture and member of horticulture department and our undergraduate and postgraduate students of horticulture who encouraged me for preparing this manuscript

I appreciate the efforts made by our colleague Dr. Hari Baksh, Dr. Bijendra Kumar Singh and Mr. Raj Pandey in collection of literature and compilation. We appreciate the efforts of M/S. New India Publishing Agency, New Delhi for printing the book in the nice form.

We don't have to express our emotional feeling towards our parents, family members and relative for their encouragement and support who have been affectionate, understanding and patient during the long hours we spent on the manuscript.

Jaunpur **(Bijendra Kumar Singh)** **(Rajaneesh Singh)**

Contents

Section A
Basic Horticulture & Pomology

Section B
Vegetable Production

Section C
Floriculture and Landscaping

Section D
Spices, Medicinal, Aromatic and Plantation Crops

Section E
Post Harvest Technology

Section A
Basic Horticulture & Pomology

1

Universities of India

1. Agriculture universities in India

A. State agriculture universities in India

1. Acharya NG Ranga Agricultural University, Guntur, Hyderabad, Telangana
2. Sri Venkateswara Veterinary University, Tirupati, Andhra Pradesh
3. Assam Agricultural University, Jorhat, Assam
4. Bihar Agricultural University, Sabour, Bhagalpur, Bihar
5. Bihar Animal Sciences University, Patna, Bihar
6. Indira Gandhi Krishi Viswavidhyalaya, Raipur, Chhattisgarh
7. Chhattisgarh Kamdhenu Visvavidyalaya, Durg, Chhattisgarh
8. Sardar Krushinagar Dantiwada Agricultural University, Dantiwada, Gujarat
9. Anand Agricultural University, Anand, Gujarat
10. Navsari Agricultural University, Navsari, Gujarat
11. Junagarh Agricultural University, Junagarh, Gujarat
12. Kamdhenu University, Gandhinagar, Gujarat
13. Chaudhary Charan Singh Haryana Agricultural University, Hisar, Haryana
14. Lala Lajpat Rai University of Veterinary & Animal Sciences, Hisar, Haryana
15. Ch. Sarwan Kumar Himachal Pradesh Krishi Viswavidyalaya, Palampur, Himachal Pradesh
16. Birsa Agricultural University, Ranchi, Jharkhand

17. Sher-e-Kashmir University of Agricultural Sciences & Technology, Shrinagar, Jammu & Kashmir
18. University of Agricultural Sciences, Bangalore, Karnataka
19. Karnataka Veterinary, Animal and Fisheries Sciences University, Bidar, Karnataka
20. University of Agricultural Sciences, Raichur, Karnataka
21. University of Agricultural Sciences, Dharwad, Karnataka
22. Kerala Agricultural University, Thrissur, Kerala
23. Kerala University of Fisheries and Ocean Studies, Panangad, Kochi, Kerala
24. Kerala Veterinary and Animal Sciences University, Pookode, Wayanand, Kerala
25. Rajmata Vijayaraje Scindia Krishi VishwaVidyalaya, Gwalior, Madhya Pradesh
26. Nanaji Deshmukh Pashu ChikitsaVishwaVidyalaya, Jabalpur, Madhya Pradesh
27. Jawaharlal Nehru Krishi Vishwa Vidyalaya, Jabalpur, Madhya Pradesh
28. Dr. Balaesahib Sawant Konkan KrishiVidyapeeth, Dapoli, Maharashtra
29. Maharastra Animal & Fisheries Sciences University, Nagpur, Maharashtra
30. Vasantrao Naik Marathwada Krishi Vidyapeeth, Parbhani, Maharashtra
31. Mahatma Phule Krishi Vidyapeeth, Rahuri, Maharashtra
32. Dr. Punjabrao Deshmukh KrishiViswaVidyalaya, Akola, Maharashtra
33. Orissa University of Agricultural & Technology, Bhubaneswar, Orissa
34. Guru Angad Dev Veterinary and Animal Sciences University, Ludhiana, Punjab
35. Punjab Agricultural University, Ludhiana, Punjab
36. Maharana Pratap University of Agriculture & Technology, Udaipur, Rajasthan
37. Swami Keshwanand Rajasthan Agricultural University, Bikaner, Rajasthan
38. Rajasthan University of Veterinary & Animal Sciences, Bikaner, Rajasthan

39. Sri Karan Narendra Agriculture University, Jobner, Rajasthan
40. Agriculture University, Kota, Rajasthan
41. Agriculture University, Jodhpur, Rajasthan
42. Tamil Nadu Agricultural University, Coimbatore, Tamil Nadu
43. Tamil Nadu Veterinary & Animal Sciences University, Chennai, Tamil Nadu
44. Tamil Nadu Fisheries University, Nagapattinam, Tamil Nadu
45. Sri PV Narsimha Rao Telangana Veterinary University, Hyderabad, Telangana
46. Professor Jayashankar Telangana State Agricultural University, Hyderabad, Telangana
47. G.B. Pant University of Agriculture & Technology, Pantnagar, Uttrakhand
48. Chandra Shekhar Azad University of Agriculture & Technology, Kanpur, Uttar Pradesh
49. Narendra Deva University of Agriculture & Technology, Ayodhya, Uttar Pradesh
50. Sardar Vallabhbhai Patel University of Agriculture & Technology, Meerut, Uttar Pradesh
51. U.P. Pt. Deen Dayal Upadhyaya Pashu Chikitsa Vigyan Vishwa Vidhyalaya Evem Go Anusandhan Sansthan, Mathura, Uttar Pradesh
52. Banda University of Agriculture and Technology, Banda, Uttar Pradesh
53. Sam Higginbottom University of Agriculture, Technology & Sciences, Allahabad, Uttar Pradesh
54. Bidhan Chandra Krishi Viswa Vidhyalaya, Mohanpur, West Bengal
55. West Bengal University of Animal & Fishery Sciences, Kolkata, West Bengal
56. Uttar Banga Krishi Viswavidhyalaya, Cooch Behar, West Bengal

B. State horticulture universities in India

1. Dr. Yashawant Singh Parmar University of Horticulture & Forestry, Solan, Nauni, Himanchal Pradesh.
2. Dr. Y.S.R. Horticulture University, Todepalligudem west Godawari, Andhra Pradesh.
3. Sri Konda Laxman Telangana State Horticulture University, Rajendranagar, Hyderabad, Telangana.
4. University of Agricultural and Horticultural Science, Navile Shimoga, Karnataka.
5. University of Horticultural Sciences, Navanagar, Bagalkot, Karnataka.
6. Veer Chandra Singh Garhwali Uttarakhand University of Horticulture & Forestry, Bharsar, Pauri Garhwal, Uttrakhand.
7. Maharana Pratap Horticultural University, Karnal, Hariyana.

C. Central agriculture universities in India

1. Central Agricultural University, Imphal, Manipur.
2. Rani Laxmi Bai Central Agricultural University, Jhansi, Uttar Pradesh.
3. Dr. Rajendra Prasad Central Agricultural University, Pusa (Samastipur), Bihar.

D. Deemed-to-be universities in India

1. Indian Agricultural Research Institute, Pusa, New Delhi.
2. Indian Veterinary Research Institute, Izatnagar, Bareilly, Uttar Pradesh.
3. National Dairy Research Institute, Karnal, Haryana.
4. Central Institute of Fisheries Education, Mumbai, Maharashtra.

E. Central universities with Agriculture Faculty/Institutes

1. Banaras Hindu University, Varanasi, Uttar Pradesh.
2. Aligarh Muslim University, Aligarh, Uttar Pradesh.
3. Vishwa Bharti, Shantiniketan, West Bengal.
4. Nagaland University, Medizipherma, Nagaland.

2

Research Institute in India

A. ICAR Agricultural Institutes

- Central Agricultural Research Institute (CARI), Post Box No. 181. Andaman Nicobar & Lakshwadeep Group of Islands, Port Blair.
- Central Arid Zone Research Institute (CAZRI), Jodhpur, (Rajasthan). Website: www.icar.org.in (Rajasthan).
- Central Avian Research Institute (CARI), Izatnagar, (Uttar Pradesh)
- Central Inland Fisheries Research Institute (CIFRI), Barrackpore (West Bengal). Web site: www.cifri.com.
- Central Institute for Cotton Research (CICR), P.O. No. 225, G.P.O., Panjari, Wardha Road, Nagpur (Maharashtra).
- Central Institute for Freshwater Aquaculture (CIFA), P.O. Kausalyaganga, Bhubaneshwar (Orissa).
- Central Institute for Research on Baffaloes (CIRB), Sirsa Road, Hisar (Haryana).
- Central Institute for Research on Cotton Technology (CIRCOT). Adenwala Road, Matunga, Mumbai (Maharashtra). Web site: education. vsnl.com/circot.
- Central Institute for Research on Goats (CIRG), P.O. Farah, Makhdoom, Mathura (Uttar Pradesh). Web site: www.cirg.res.in
- Central Institute of Agricultural Engineering (CIAE), Nabibagh, Berasia Road, Bhopal (HP).
- Central Institute of Brackishwater Aquaculture (CIBA), 141, Marshal's Road, Egmore, Chennai (Tamil Nadu). Web Site: www.ciba.tn.nic.in
- Central Institute of Fisheries Education (CIFE), Jaiprakash Road, Seven Bungalows, Versova, Mumbai (Maharashtra). Web Site: www.icar.org.in/cife/index.html

- Central Institute of Fisheries Technology (CIFT), Willingdon Island, P.O. Matsyapuri, Cochin (Kerala).
- Central Marine Fisheries Research Institute (CMFRI), PO No. 1003, Emakulam, Cochin (Kerala), Web site www.cmfneom.
- Central Research Institute for Dryland Agriculture (CRIDA), Santos Nagar, PO. Saidabad, Hyderabad (Telangana).
- Central Research Institute for Jute & Allied Fibers (CRIJAF), 24 Paraganas, Barrackpore (West Bengal).
- Central Rice Research Institute (CRRI) Cuttak (Orissa).
- Central Sheep & Wool Research Institute (CSWRI), Avikanagar, Tehsil Malpura, Tonk (Rajasthan).
- Central Soil & Water Conservation Research & Training Institute (CSWCRTI), 212, Kaulagarh Road, Dehradun (Uttarakhand).
- Central Soil Salinity Research Institute (CSSRI), ZarifaFarm, Kachwa Road, Karnal (Haryana), Web site: www.cssri.org
- Central Tabacco Research Institute (CTRI), Rajahmundry (Andhra Pradesh).
- ICAR Research Complex for Eastern Region, Walmi Complex, Phulwari Shariff, Sreekartiyam, Patna (Bihar).
- ICAR Research Complex for Goa (ICARRCG), Ela Old (Goa). Web site: icargoa.res.in
- ICAR Research Complex for NEH Region (ICARRCNEHR), Umroi Road, Barapani (Meghalaya)
- Indian Agricultural Research Institute (1ARI), Pusa, Dr. K.S. Krishnan Marg, (New Delhi), Web site: www.iari.re.in
- Indian Agricultural Statistics Research Institute (IASRI), Library Avenue, Pusa, (New Delhi). Web site: iasri.res.in
- Indian Grassland & Fodder Research Institute (IGFRI), Pahuj Dam, Jhansi-Gwalior Road, Jhansi, (Uttar Pradesh). Web site: www.iasri.res.in
- Indian Institute of Pulses Research (IIPR), Kalyanpur, Kanpur (Uttar Pradesh). Web site: http://iipr.up.nic.in
- Indian Institute of Soil Sciences (IISS), Nabi Bagh, Berasia Road, Bhopal (Madhya Pradesh).

- Indian Institute of Sugarcane Research (IISR), Raibareli Road, P.O. Dilkusha, Lucknow (Uttar Pradesh).
- Indian Lac Research Institute (ILRI), Namkum, Ranchi (Jharkhand).
- Indian Veterinary Research Institute (IVRI), Izatnagar (Uttar Pradesh).
- National Institute for Research on Jute Allied Fibres Technology (NIRJAFT), 12, Regent Park, Calcutta (West Bengal).
- National Academy of Agricultural Research Management (NAARM), Rajendranagar, Hyderabad (Telangana).
- National Dairy Research Institute (NDRI), Karnal (Haryana).
- National Institute of Animal Nutrition & Physiology (NIANP), NDRI Campus, Adugodi, Bangalore (Karnataka). Web site: http://www.nianp.res.in
- Sugarcane Breeding Institute (SBI), Coimbatore (Tamil Nadu).
- Vivekanand Parvatiya Krishi Anusandhan Shala (VPKAS), Almora (Uttar Pradesh). Web site: http://vpkas.nic.in

B. ICAR Horticultural Institutes

Horticultural Institutes

- Central Institute of Horticulture (CIH), Medziphema, Nagaland
- Institute of Horticulture Technology (IHT), New Delhi
- Indian Institute of Horticultural Research (IIHR), Hessarghatta, Bangalore, Karnataka
- National Bureau of Plant Genetic Resource Center (NBPGR), New Delhi
- National Horticultural Board (NHB) established in 1984, HQ in Gurgaon, Haryana
- National Horticultural Mission (NHM)

Pomological Research Institutes

- Central Institute of Subtropical Horticulture (CISH), Lucknow, Uttar Pradesh
- Central Institute of Temperate Horticulture (CITH), Srinagar, Jammu and Kashmir
- Central Arid Zone Research Institute (CAZRI), Jodhpur, Rajasthan

- Central Institute of Arid Horticulture (CIAH) (Previously, NRC for Arid Horticulture), Bikaner, Rajasthan
- National Research Centre for Banana (NRCB), Trichy, Tamil Nadu
- National Research Centre for Citrus (NRCC), Nagpur, Maharashtra
- National Research Centre for Grapes (NRCG), Pune, Maharashtra
- National Research Centre for Pomegranate (NRCP), Solapur, Maharashtra
- National Research Centre for Litchi (NRCL), Muzaffarpur, Bihar
- National Research Centre for Makhana (NRCM), Darbhanga, Bihar

Olericultural Research Institutes

- Indian Institute of Vegetable Research (IIVR), Varanasi, Uttar Pradesh
- Central Tuber Crops Research Institute (CTCRI), Sreekariyam, Thiruvananthapuram, Kerala
- Central Potato Research Institute (CPRI), Shimla, Himachal Pradesh
- National Horticultural Research and Development Foundation (NHRDF), Nasik, Maharashtra
- Directorate on Onion and Garlic Research (DOGR), Nasik, Maharashtra (Previously known as National Research Center for Onion and Garlic (NRCOG)
- Directorate of Mushroom Research (DMR), Solan, Himachal Pradesh (Previously known as National Research Centre for Mushroom (NRCM))

Floricultural Research Institutes

- Directorate of Floricultural Research (DFR), IARI, (New Delhi)
- National Research Centre for Orchids (NRCO), Pakyong, Gangtok, Sikkim
- National Botanical Research Institute (NBRI), Lucknow, Uttar Pradesh
- Institute of Himalayan Bio-resource Technology (IHBT), Palampur, Himachal Pradesh

Plantation Crops Research Institutes

- Central Plantation Crops Research Institute (CPCRI), Kasargod, Kerala
- Central Coffee Research Institute (CCRI), Chikmagalur, Karnataka

- Directorate of Cashew Research (DCR), Puttur, Karnataka (Previously known as National Research Centre for Cashew (NRCC)
- Directorate of Oil Palm Research (DOPR), Pedavegi, Eluru, Andhra Pradesh (Previously known as National Research Centre for Oil Palm (NRCOP)
- United Planters' Association of Southern India (UPASI), Glenview, Conoor, Nilgiris, Tamil Nadu
- UPASI Tea Research Foundation (UPASI TRF), Valparai, Coimbatore, Tamil Nadu
- Tea Research Institute (TRI), Nirar Dam, Valparai, Tamil Nadu
- Central Arecanut and Cocoa Marketing and Processing Co-operative Limited (CAMPCO), Mangalore, Karnataka
- Directorate of Cashew and Coconut Development (DCCD), Cochin, Kerala
- Directorate of Arecanut and Spices Development (DASD), Calicut, Kerala

Spices Research Institutes

- Indian Institute of Spices Research (IISR) (Previously, NRC on Spices), Calicut, Kerala.
- National Research Centre for Seed Spices (NRCSS), Tabiji, Ajmer, Rajasthan
- Indian Cardamom Research Institute (ICRI), Myladumpara, Idukki, Kerala

Medicinal and Aromatic Plants Research Centres

- Central Institute of Medicinal and Aromatic Plants (CIMAP), Lucknow, Uttar Pradesh
- Directorate of Medicinal and Aromatic Plants Research (DMAPR), Anand, Gujarat (Previously known as National Research Center for Medicinal and Aromatic Plants (NRCMAP).

Post Harvest Research Centers in India

- Indian Institute of Crop Processing Technology (IICPT), Thanjavur, Tamil Nadu
- Fruit Preservation and Canning Institute (FPCI), Lucknow, Uttar Pradesh

- Central Institute of Post Harvest Engineering and Technology (CIPHET), Ludhiana, Punjab
- Central Food Technological Research Institute (CFTRI), Mysore, Karnataka
- Central Food Laboratory (CFL), Kolkata, West Bengal
- Food Research and Standardization Laboratory (FRSL), Ghaziabad, Uttar Pradesh
- Defence Food Research Laboratory (DFRL), Mysore, Karnataka
- National Institute of Food Technology Entrepreneurship and Management (NIFTEM), Kundli, Haryana
- National Agricultural Cooperative Marketing Federation of India Ltd.(NAFED) New Delhi
- Regional Research Laboratory (RRL) Jammu
- Bhabha Atomic Research Centre (BARC), Trombay, Bombay
- Public Health Laboratory (PHL), Pune, Maharashtra
- Agricultural and Processed Food Products Export Development Authority (APEDA), New Delhi

C. ICAR National Research Centers (NRC) for Agriculture

- National Centre for Agricultural Economics & Policy Research (NCAEPR), IASRI Campus, Library Avenue, Pusa, New Delhi.
- National Centre for Integrated Pest Management (NCIPM), Lal Bahadur Shastri Bhawan, Wing L-1 & M-1, Block F, IARI Campus, New Delhi.
- National Research Centre for Agroforestry (NRCAF), IGFRI Campus, Pahuj Dam, Jhansi-Gwalior Road, Jhansi (U.P).
- National Research Centre for Camel (NRCC), Jorbeer, P.O. Box. No. 7, P.O. Box No. 07, Bikaner (Rajasthan).
- National Research Centre for DNA Fingerprinting (NRCDF), NBPGR, IARI Complex, Pusa, (New Delhi).
- National Research Centre for Equines (NRCE), Sirsa Road, Hisar (Haryana).
- National Research Centre for Groundnut (NRCG), Ivanagar Road, P.B. No. 5, Junagarh (Gujarat).

- National Research Centre for Meat & Meat Products (NRCMMP), PD Poultry Science, APAU Campus, Rajendranagar, Hyderabad (Telangana).
- National Research Centre for Mithun (NRCM), ICAR Research Complex, Jharnapani, Medziphema (Nagaland).
- National Research Centre for Rapeseed and Mustard (NRCRM), Sewar, Bharatpur (Rajasthan).
- National Research Centre for Weed Science (NRCWS), P.B. No. 17, Maharajpur, Adhartal, Jabalpur (Madhya Pradesh).
- National Research Centre for Women in Agriculure (NRCWA), 1199, Jagamana, P.O. Khandagiri, Bhubaneswar (Orissa).
- National Research Centre for Yak (NRCY), Dirang, West Kameg, (Arunachal Pradesh).
- National Research Centre on Coldwater Fisheries (NRCCWF), "Saurabh Cottage". ThandiSarak, Bhimtal (Uttarakhand).
- National Research Centre on Plant Biotechnology (NRCPB), IARI Campus, Pusa, (New Delhi)
- National Research Centre on Sorghum (NRCS), Rajendranagar, Telangana, (Telangana).
- National Research Centre on Soybean, (NRCS) Khandwa Road, Indore (Madhya Pradesh).

D. National Bureaus

- National Bureau of Fish Genetic Resources (NBFGR), 351/28, Radha Swami Bhawan, Duriyapur, P.O. Rajendranagar, Lucknow (Uttar Pradesh).
- National Bureau of Animal Genetic Resources (NBAGR), P.B. No. 129, Karnal (Haryana). Web site: www.icar.org.in/nbagr/nbagr.html
- National Bureau of Plant & Genetic Resources (NBPGR), Pusa Complex, Inder Puri (New Delhi). Website: nbpgr.delhi.nic.in
- National Bureau of Soil Survey & Land Use Planning (NBSSLUP), Shankar Nagar, Amravati Road, Nagpur (Maharashtra).
- National Bureau of Agricultural Important Micro-Organisms (NBAIMO), Old NBPGR Building, Pusa, Campus, New Delhi.

E.Project Directorate for Agriculture

- Water Technology Centre for Eastern Region (WTCER), Chandra Shekharpur, Near NALCO Nagar, Bhubaneswar (Orissa).Website: www.wtcer.stpbh.soft.net (now known as Indian Institute of Water Management)
- Project Directorate of Biological Control (PDBC) P.O. Box 2491, H.A. Farm, Post, Balgalore (Karnataka).
- Project Directorate of Cropping System Research (PDCSR), Modipuram, Meerut (Uttar Pradesh).
- Project Directorate of Oilseed Research (PDOR), Rajendranagar, Hyderabad (Telangana). Web site: www.dor-icar.org
- Project Directorate of Rice Research (PDRR), Rajendranagar, Hyderabad (Telangana). Web site www.drrindia.org
- Project Directorate of Water Management Research (PDWMR), Walmi Complex. P.O., Phulwari Sharif, Patna (Bihar).
- Project Directorate of Wheat Research (PDWR), P.O. Box. 158, Kunjpura Road. Karnal-132 001 (Haryana). Web site : www.nic.in/icarfdwt/dwrmain.htm (now known as Indian Institute of Wheat and Barely)
- Project Directorate on Cattle (PDC), PH-7, Pallavpuram, Phase II, Modipuram. Meerut (Uttar Pradesh). Web site : pdcattel.up.nic.in
- Project Directorate on Maize (PDM), Cummings Laboratory, IARI Campus, Pusa, New Delhi.
- Project Directorate on Poultry (PDP), APAU Campus, Rajendranagar, Hyderabad, (Telangana).

F. Important Board in Horticulture

- Coconut Development Board, Ministry of Agriculture, Govt of India, Krishi Bhavan, SRVHS Road, Kochi (Kerala).
- Tea Board of India, Ministry of Commerce, 14, B.T.M. Sarani, Kolkata (WB).
- Coffee Board of India, Chikmagalur, Karnataka
- Çashew Export Promotion Council of India (CEPCI), Ernakulam, Kerala.
- Spices Board of India, Ministry of Commerce, Government of India, Sugandha Bhavan, Cochin (Kerala).

- National Medicinal Plants Board (NMPB), New Delhi.
- National Horticultural Board, Ministry of Agriculture, Govt. of India, 85, Institutional area, Sector-18, Gurgaon (Haryana).
- National Bee Board, Ministry of Agriculture, Govt. of India, W-42, Greater Kailash, Part II, New Delhi.

G. Horticultural Societies

- Agri-horticultural Society of India (AHSI), 1820, Kolkata, India
- Indian Society of Vegetable Science (ISVS), 1993, New Delhi, India
- Indian Society of Ornamental Horticulture (ISOH), 1990, New Delhi, India
- International Society of Horticultural Science (ISHS), Leuvien, Belgium: 1959
- American Society of Horticultural Science (ASHS), 1903, Duke Street Alexandria, United States of America
- Royal Horticultural Society (RHS) was founded in London, England, 1804

H. International Horticultural Research Centre

- Global Horticulture Initiative (GHI), Rome, Italy.
- Horticulture Research International (HRI), Wellesbourne, United Kingdom.
- International Network for the Improvement of Banana and Plantain (INIBAP), Montpellier, France.
- Biodiversity International (Previously known as International Plant Genetic Resources Institute (IPGRI)), Rome, Italy.
- World Vegetable Center (WVC), Taiwan (Previous ly known as Asian Vegetable Research and Development Center, AVRDC).
- International Potato Centre (CIP), Peru.
- International Registration Authority for Rose (IRAS), USA.
- International Registration Authority for Bougainvillea (IRAB), New Delhi.
- International Flower Market (IFM), Alsmeer, Netherland.
- International Flower Auction Centre (IFAC), Bangalore, Karnataka.

- International Cut Flower Grower Association, (ICFGA) USA.
- International American Spice Trade Association (IASTA), Washington, D.C., USA.
- Royal New Zealand Institute of Horticulture (RNZIH), Canterbury, New Zealsnd.

I. All India Coordinated Research Project (AICRP)

- AICRP on Tropical Fruits, Bangalore, Karnataka
- AICRP on Sub-Tropical Fruits, Lucknow, Uttar Pradesh
- AICRP on Arid Zone Fruits, Bikaner, Rajasthan
- AICRP on Vegetables including NSP vegetable, Varanasi, Uttar Pradesh
- AICRP on Tuber Crops, Thiruvananthapuram, Kerala
- AICRP on Potato, Shimla, Himachal Pradesh
- AICRP on Mushroom, Solan, Himachal Pradesh
- AICRP on Floriculture, New Delhi
- AICRP on Cashew, Puttur, Karnataka
- AICRP on Palms, Kasaragod, Kerala
- AICRP on Spices, Calicut, Kerala
- AICRP on Medicinal and Aromatic Plants including Betel vine, Anand, Gujarat

J. Statutory Provisions for Quality Control in India

- Prevention of Food Adulteration (PFA) Act: 1954
- Fruits Product Order (FPO) 1955
- AGMARK: Agricultural Produce (Grading and Marketing) Act: 1937
- Central Agmark Laboratory is Located at Nagpur, Maharashtra
- Export (Quality Control and Inspection) Act: 1963
- Export (Quality Control and Inspection) Rules: 1964
- The Consumer Protection Act: 1986
- Food Safety and Quality Control Act: 2001
- Food Safety and Standardization Authority (FSSAI) act: 2005

K. Central Food Laboratories

- Central Food Laboratory (CFL), Kolkata, West Bengal
- Food Research and Standardization Laboratory (FRSL), Ghaziabad, Uttar Pradesh
- Public Health Laboratory (PHL), Pune
- Central Food Technological Research Institute (CFTRI) Mysore, Karnataka
- Defence Food Research Laboratory (DFRL), Mysore, Karnataka
- Central Institute of Post Harvest Engineering Technology (CIPHET), Ludhiana, Punjab
- National Institute of Food Technology Entrepreneurship and Management (NIFTEM) Kundli, Haryana
- Indian Institute of Crop Processing Technology (IICPT), Thanjavur, Tamil Nadu
- National Agricultural Cooperative Marketing Fedration of India Ltd. (NAFED), New Delhi.

3

Journals, Magazines and Books

A. Research journal

1. ***Haryana Journal of Horticultural Science*** **(Quarterly)**- Society of Haryana, Department of Vegetable Crops, CCS, HAU, Hisar (Haryana).
2. ***Hort. Science*** **(Annualy)**- American Society for Horticultural Sciences 600, Camerun Street, Alexandria, USA
3. ***Horticulture Research*** **(Half yearly)**- Scottish Academic Press, 25, Perth Street, Edinburgh.
4. ***Indian Journal of Horticulture*** **(Quarterly)**- Horticulture Society of India, Division of Fruits and Horticultural Technology, IARI, Pusa, (New Delhi).
5. ***Journal of Applied Horticulture*** **(Half yearly)**- Managing Editor, Journal of Applied Horticulture, A-859, Indiranagar, Lucknow (UP).
6. ***Progressive Horticulture*** **(Quarterly)**- Hill Horticulture Development Board, Horticultural Experiment and Training Centre, Chaubattia, Ranikhet, (Uttarakhand).
7. ***Scientia Horticulture*** **(Annualy)**- Elsevier Science P.O. Box 945, New York,
8. ***The Horticultural Journal*** **(Quarterly)**- Society of Advancement of Horticulture, Faculty of Horticulture, BCKV, Mohanpur, (WB).
9. ***The Journal of Horticultural Sciences*** **(Annualy)**- Headley Brothers Ltd. The Invicta Press, Ashford, Kent, England.
10. ***Vegetable Science*** **(Half yearly)**- Indian Institute of Vegetable Research (IIVR), P.O. Jakhani (Shahanshapur), Varanasi- (UP).

B. Magazines

1. Indian Horticulture - ICAR, New Delhi.
2. Phal-Phool - ICAR, New Delhi.
3. Rose World – Lucknow.
4. Kisan World – Chennai.
5. Krishi Chayanika- ICAR, New Delhi
6. Popular Kheti -Rajasthan

C. Important books

1. **Advances in Horticulture-** K.L. Chaddha, Malhotra Publishing House, New Delhi.
2. **Basic Horticulture-** Jitendra Singh, Kalyani Publishing, New Delhi.
3. **Basics of Horticulture (3rd Revised and Enlarged Edition)**, K.V. Peter, New India Publishing Agency, (New Delhi).
4. **Beautiful Climbers-** B.P. Pal, ICAR, (New Delhi).
5. **Breeding of Horticultural Crops: Principles and Practices (3rd Revised and Enlarged Edition)**, New India Publishing Agency, New Delhi
6. **Commercial Flower-**T.K. Bose and L.P. Yadav, Naya Prokash, Kolkata, (West Bengal)
7. **Cultivation of Spices Crops-**A.A Farooqi, B.S. Sreeramu and Srinivasappa, University Press, Hyderabad.
8. **Floriculture at Glance-**Desh Raj, Kalyani Publishers, Ludhiana.
9. **Floriculture in India-**G.S. Randhawa and A. Mukkopadhyay, Allied Publishers, New Delhi.
10. **Flowering Shrubs in India-**S.L. Jindal, Publications Division
11. **Handbook of Horticulture-** K.L. Chaddha, ICAR, (New Delhi).
12. **Horticulture at a Glance-** A.S. Salaria and B. Salaria, Jain Brothers, New Delhi.
13. **Horticulture Science books -** Jules Jenick, W.H. Freeman
14. **Introduction to Spices, Plantation Crops, Medicinal and Aromatic Plants-**N. Kumar, Oxford & IBH Pub. Co. Pvt. Ltd. New Delhi.

15. **Introductory Ornamental Horticulture**-J.S. Arora- Kalyani Publishers, Ludhiana.

16. **Major Spices of India-Crop Management and Post Harvest Technology** - J.S. Pruthi, ICAR, New Delhi.

17. **Minor Spices and Condiments - Crop Management and Post Harvest Technology**- J.S. Pruthi, ICAR, New Delhi.

18. **Ornamental Horticulture** - V. Swaroop, Macmillian India Ltd., New Delhi.

19. **Ornamental Horticulture in India** - K.L. Chadhha and B. Chaudhary, ICAR, New Delhi.

20. **Post Harvest Handing of Fruits and Vegetables**- Pantastico

21. **Preservation of Fruits and Vegetables**-R.P. Shrivastav and Sanjeev Kumar, IBDC Pub. Lucknow.

22. **Presrvation of Fruits and Vegetables** - G.L. Siddappa, ICAR, New Delhi.

23. **Glimpses of Post Harvest Technology**- U.S. Meena, New Vishal Pub., New Delhi.

24. **Tropical Horticulture** - T.K. Bose, Naya Prokash, Kolkata, West Bengal.

25. **Subtropical Horticulture** - T.K. Bose, Naya Prokash, Kolkata, West Bengal.

26. **Temperate Horticulture** - T.K. Bose, Naya Prokash, Kolkata, West Bengal.

27. **Text Book of Floriculture and Landscaping** - A.K. Singh and A.S. Sisodia- NIPA Publication, New Delhi.

28. **Text Book of Pomology** - T.K. Chattopadhyay, Kalyani Publishers, New Delhi.

29. **Vegetable Crops** - T.K. Bose, Naya Prokash, Kolkata, West Bengal.

30. **Vegetable Crops of India** - K.S. Yawalkar, Agrihorti Pub. Co., Nagpur, Maharastra.

4

Tools and Equipments

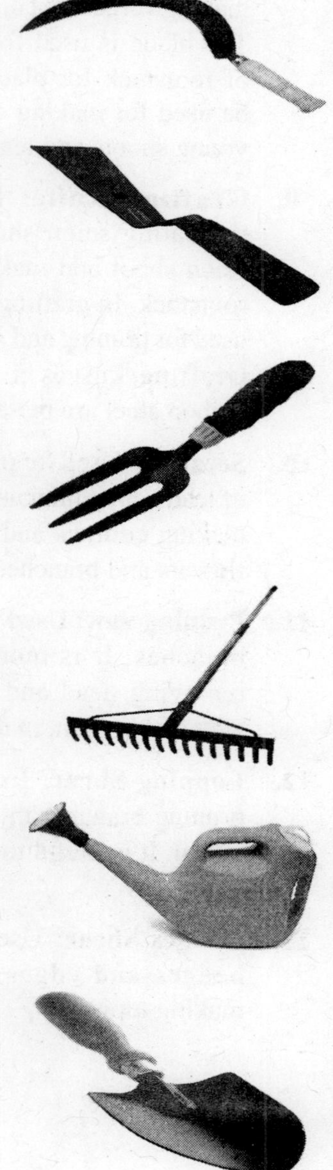

1. **Sickle:** Used for cutting grass, fodder and harvesting crop. Sickle made of steel is better and lasts longer.
2. **Khurpi:** An useful tool for weeding in vcgetable garden and towers teds. It is used for lifting seedlings and plants from the nursery. Useful tool tor transplanting and planting plants, The shape and the size determines the nature of work which can be taken with khurpi.
3. **Hand Hoe:** Effective and efficient tool for weeding in beds and gardens. It can be used for preparing basins, beds and seedbeds. Long wooden handle enables to work without bending. Ideal steel for hoe is high carbon steel.
4. **Rake:** Ideal tool for levelling land and collecting weed. It can also be used for breaking clods. Rake made up of high carbon steel lasts long. A good rake should have teeth's.
5. **Watering Can:** Used for irrigating potted plants when fitted with rose, it can be used for irrigating seed beds. They are available various capacities.
6. **Trowel:** Useful handy tool for lifting young seedling from nursery beds. It helps in lifting seedling without much injury to roots. It is available in market in different shape and size.

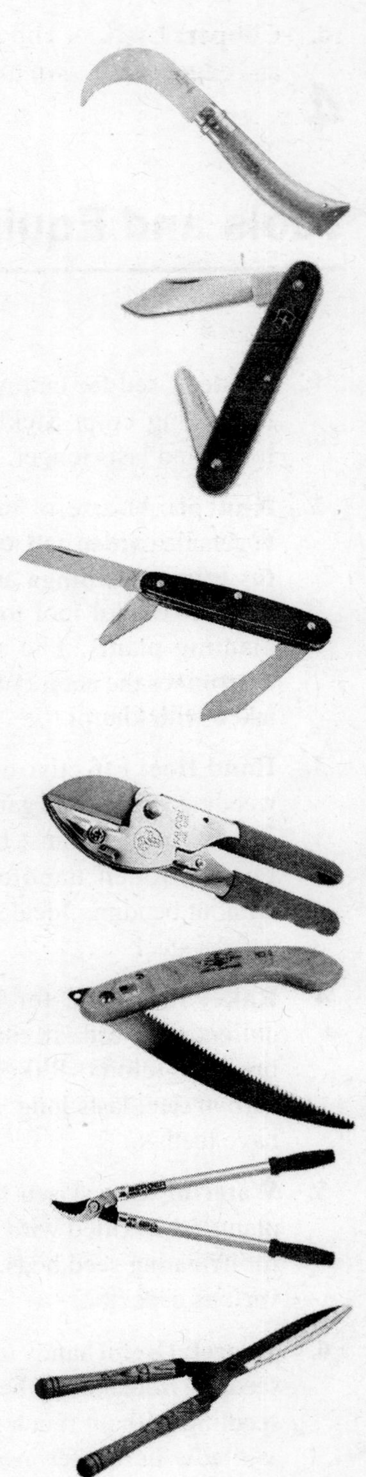

7. **Pruning Knife**: Used for pruning branches and suckers. It can be used for grafting purposes too. The edge of knife remains sharp for a longer period if it is made from high carbon steel.

8. **Budding Knife**: Used in budding operations. Curved points help making vertical cut on the stock plant. The blade is used for opening bank of rootstock for placing bud. It can be used for making cutting, pruning young shoots and tender branches.

9. **Grafting Knife:** It is used for separating scion shoot, defoliating scion shoot and making incision on rootstock. In grafting operation, also used for pruning and making cuttings. Grafting knives made from high carbon steel are better

10. **Secateur:** Used for pruning branches of lead pencil thickness. It is used for making cuttings and removing dead flowers and branches.

11. **Pruning Saw:** Used for cutting thick branches. It is more effective for removing dead and thick branches which over 6cm, in diameter.

12. **Lopping Shear:** Excellent tool for pruning branches thicker than a lead pencil. It is useful tool for hardwood cutting.

13. **Hedges Shear:** Used for trimming hedges and edges. Also used in making topiary

14. Clipper: Used for clipping soft grass of lawn and edge where lawn mower not cover.

15. Fork: Used for lifting manures and plant. Useful garden tool in nursery preparation. It is also useful for shifting compost and garden rubbish.

16. Pole Shear: Useful for pruning branches of outer periphery of the tree and are 3-4 meter high. The branch is fastened by opening the blade. The blade cuts the branches when rope or chain is pulled.

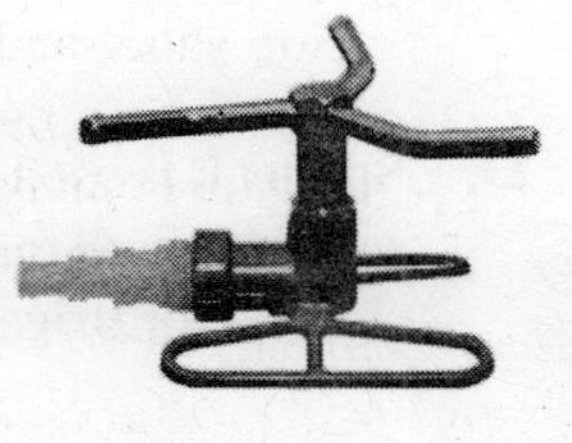

17. Sprinkler: Useful for irrigating lawn and kitchen garden. Water comes out as spray. There is economy of water as well and it also covers larger area in shorter time.

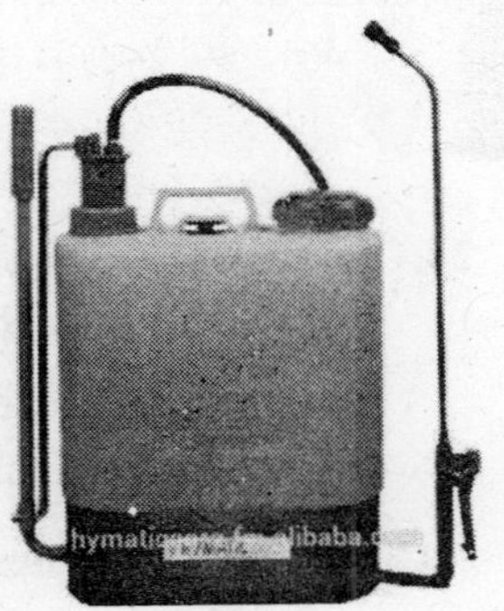

18. Knap-Sack-Sprayer: It is back mounted. Sprayers are useful for spraying insecticide, fungicide, or herbicides. It is also used for foliar application of fertilizers. It should have a fine nozzle for uniform spray. Sprayers are available in the market in different capacities. They may be foot operated or hand compressed sprayers.

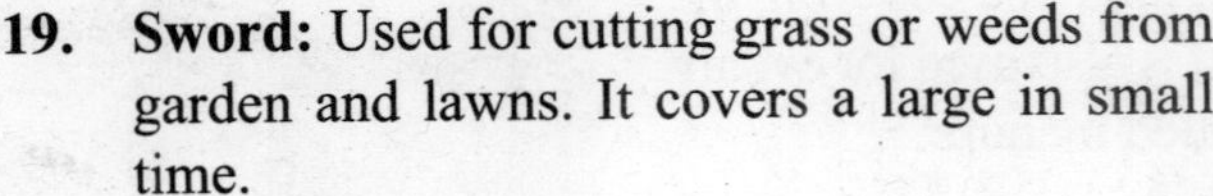

19. Sword: Used for cutting grass or weeds from garden and lawns. It covers a large in small time.

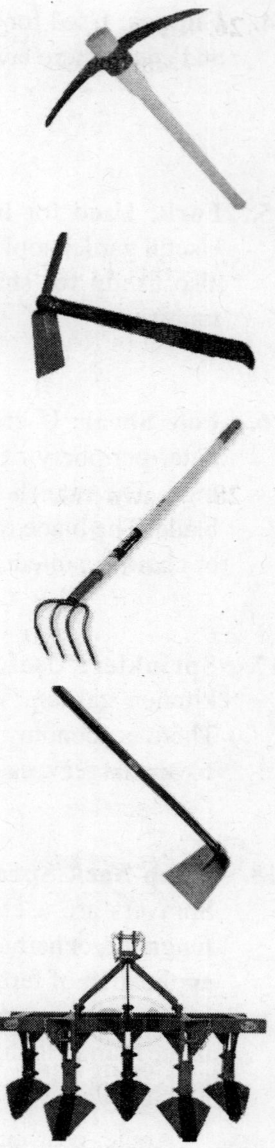

20. **Pick Axe.** It is made of carbon steel. Pick axe have two edges with provision of axial hole for attachment with handle. One edge of pick axe is pointed and another is broadened. It is used for digging hard, compact and stony soils.

21. **Kudali.** It can also be said as single pick axe. It is used for digging compact soil.

22. **Pronged Hoes,** It is having two, three or four arms. It is used for digging of hard stony sol and also for digging tuber crops like potato, tunipete. Besides, it is useful for mixing manure.

23. **Spade.** It is used for lifting and turning the soil. Also used for digging the pit, preparing channel for irrigation and drainage.

24. **Furrow Opener.** It is used for opening narrow and shallow furrow after sowing seeds in the nursery.

25. **Hand Leveller.** It is used in small beds and nursery for levelling land and covering the seeds after sowing. It is also used for evenly distributing the applied manure.

26. **Axe.** It is used for felling and cutting branches.

27. **Tree Pruner.** It is used for punning trees in which the high up branches remain out of reach from the ground level.

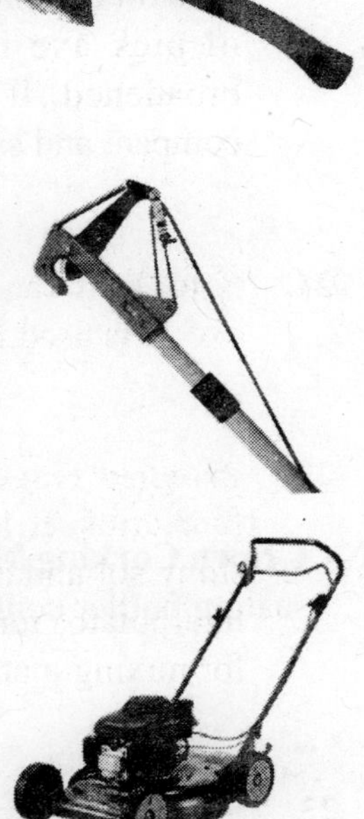

28. **Lawn Mower-** It is used for cutting lawn grass from ground layer. It is found in the electrical or manuals.

29. **Knife-** It is used for cutting of fruits and vegetable.

i. **Peeling knife-** Used for removing outer covering of fruit and vegetables.

ii. **Pitting knife-** Removal of stone from the fruit is termed as pitting and it is practiced using pitting knife.

iii. **Coring knife-** Some fruit have hard core inside them which is known as core eg. Apple, pear, pineapple etc. The core of the fruit is removed by coring knife.

iv. **Cutting knife-** Used for cutting fruit and vegetable.

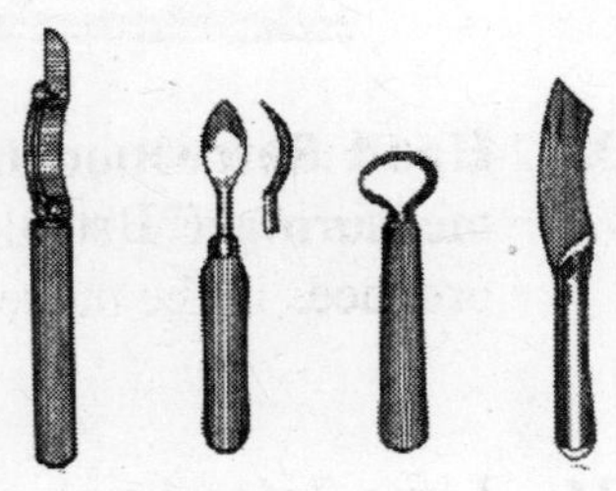

30. **Can Sealer-** It is used for sealing of can.

31. **Crown Corking Machine-** It is used for sealing bottler containing beverages.

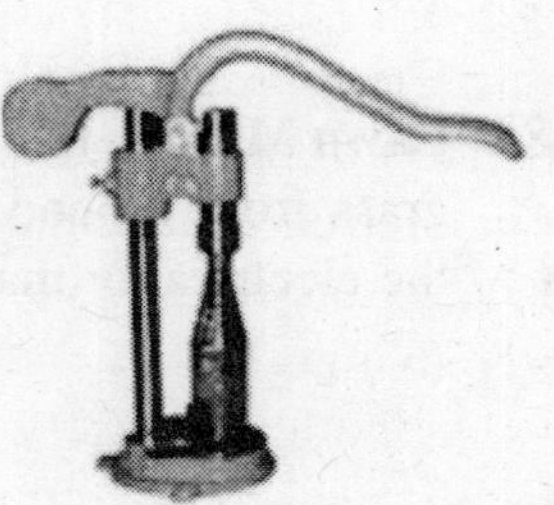

32. **Jelmeter-** It is used for measuring pectin content of fruit extract.

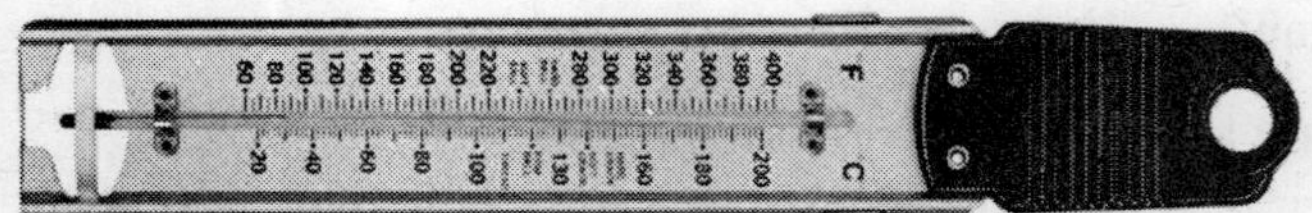

33. **Hand Refrectometer-** It is used for measuring (°Brix) sugar contain of produce.

34. **Lime Juice Squeezer-** It is used for obtaining juice from individual fruit of lime.

35. **Rootex-** This is rooting hormone (synthetic IBA) which is available in the market trade name rootex (1, 2 and 3 No. which is used in hard, soft and herbaceous cutting respectively)

5

Potting, Shifting and Filling of Pots

Purpose- Burnt porous clay or cement or plastic pots are commonly used for various purposes in horticulture. The pots are used in propagation or plants, cultivation of ornamental plants and flowering annuals. Growing of plants in pots in referred as pot-culture. In pot-culture, operations such as potting, shifting and filling of pots are very important. Potting refers to planting of plants in pots containing soil mixture, while shifting means removal of plants from pots and planting them again in the same or different pot. Plants growing for more than one year in the same pot need shifting. The time for potting and shifting of plants is monsoon season (July to Sept.) and during spring (Feb-March).

Materials and Equipments- Flower pots, crocks (broken pieces of pots), potting mixture, khurpi and watering can (with fine nozzle).

A. FILLING OF POTS

Procedure- Select well-baked and sound clay pots of the proper size (the size will depend on age of the plant and purpose for which the pots are filled). Wash the pot, both inside and outside, with clean water. Place a large crock on the drainage hole (Plate 1-1). Now put several smaller crocks to a depth of 5 to 10 cm. dependent upon the size of the pot. Put 1.2 to 2.5 cm layer of coarse sand or coconut fiber. This necessary to check the washing away of the fine soil particles. Fill the pot with soil mixture or compost. When half full, press the mixture firmly. Continue filling up to the rim of the pot. Press the potting mixture again, and finally fill and press mixture to a point where a room of 2 to 3 cm. is left for holding water.

B. POTTING

Procedure - The pots filed as described above can be used for sowing seeds, potting of plants or planting cuttings. For potting of plants scoop out a hole in the centre of the field pot. Keep the plant in the centre of the hole with roots well distributed in all directions. Throw potting mixture all round the plant and press it firmly and uniformly. Care should be taken that planting is not done too deep (Plate 1-4). Irrigate the plant with watering can fitted with a fine rose. Place the potted plant in a cool shady place for establishment.

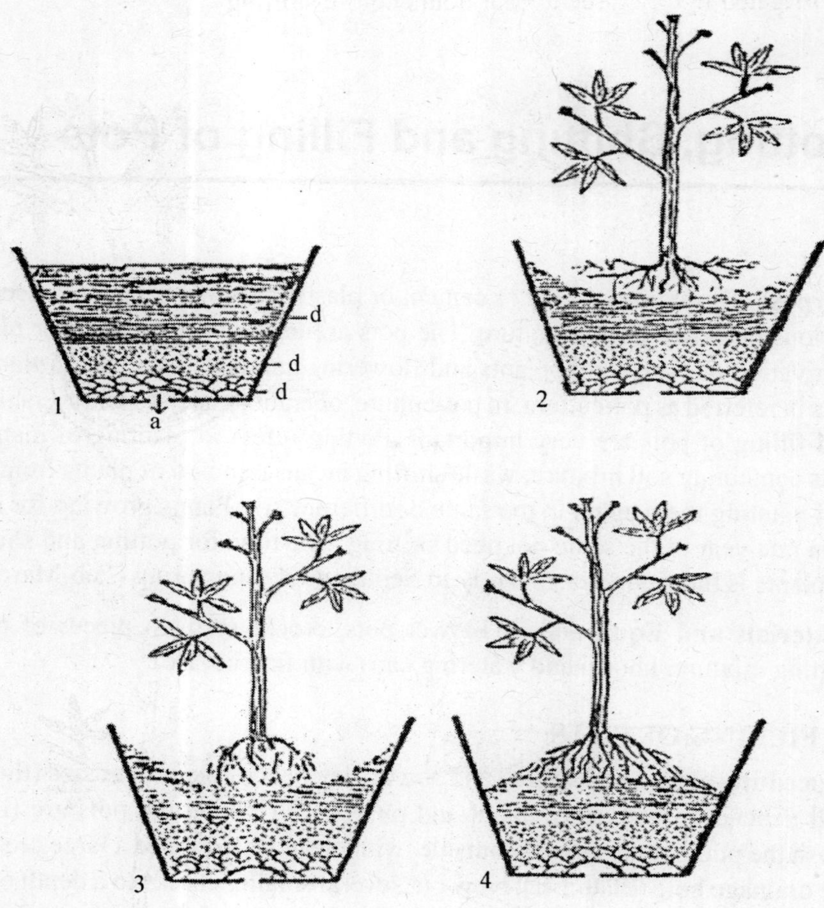

1. (a) drain hole, (b) crock, (c) sand, (d) soil mixture
2. & 3. Incorrect method of planting
4. Correct method of planting

Plate-1: Steps Showing Potting of Plants

C. SHIFTING OF PLANTS

Procedure - Each potted plant needs shifting every year to avoid pot bound condition. This operation should be done during the rainy season, In order to remove plant from the pot. It is held with right hand second and third finger and the thumb along the side of the pot. The pot is then turned down as shown in plate 2; and then taped gently on the edge of bench or on a pot to loosen the ball of earth. The ball of earth is then taken cut from the pot carefully, before putting the plant in a new pot the lower fine roots and some soil is removed. The various steps involved are illustrated in Plate-2. If necessary me pots be irrigated lightly three to four hours before shifting

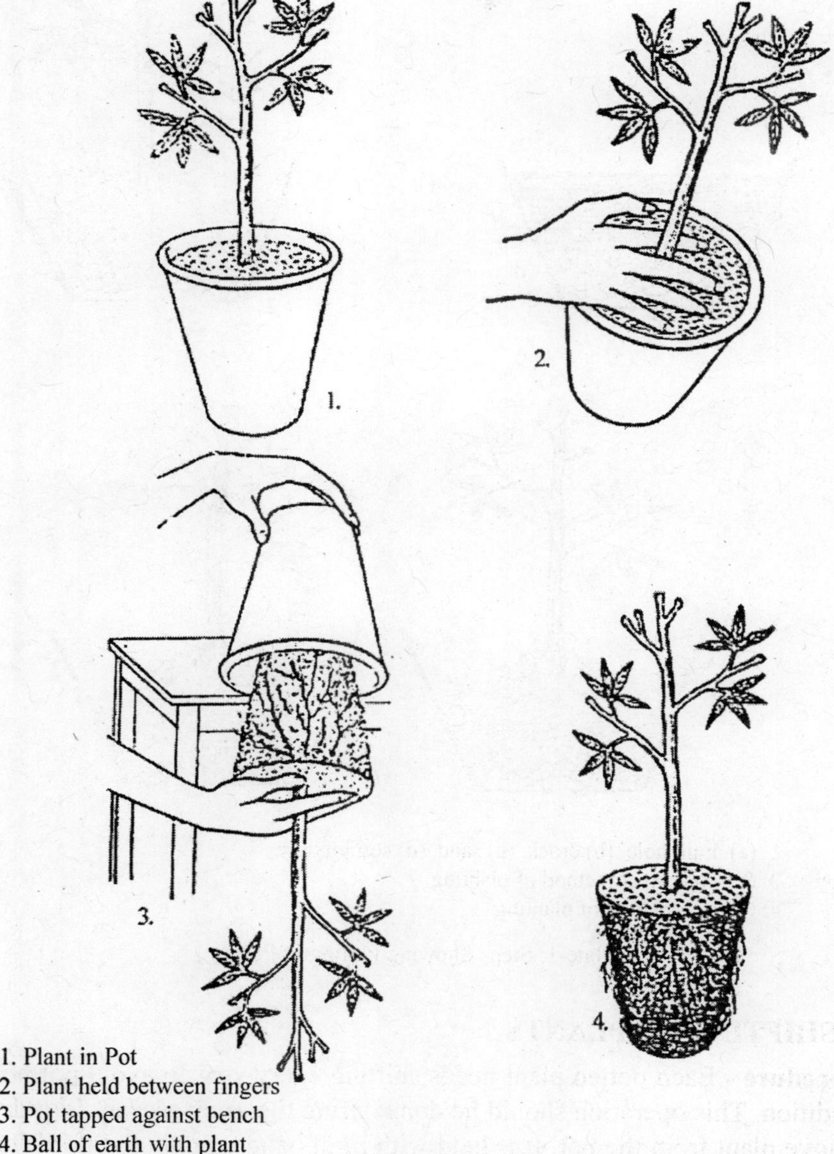

Plate 2: Shifting of Potted Plant

6

Orchard Planning

Purpose- There are several planting plans or systems which can be adopted for planting orchard. The following points need to be considered before choosing any system of planting.

(1) It should accommodate maximum number of plants per unit area.

(2) It should allow sufficient space for the development of each tree.

(3) Orchard operation such as ploughing, dusting are easy to perform.

The important orchard planting plans are as follow:

1. Square
2. Rectangular
3. Triangular
4. Hexagonal
5. Quincunx
6. Contour

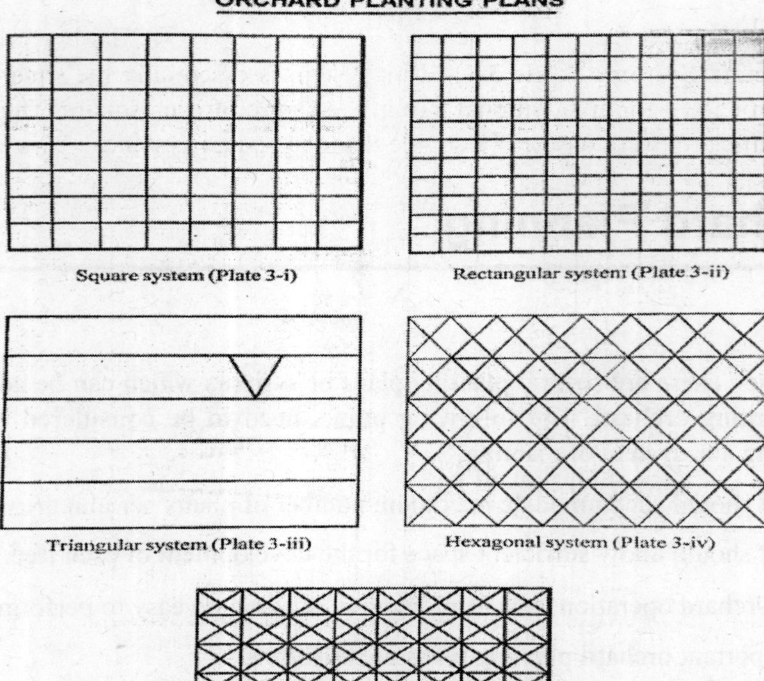

Square system (Plate 3-i)

Rectangular system (Plate 3-ii)

Triangular system (Plate 3-iii)

Hexagonal system (Plate 3-iv)

Quincunx system (Plate 3-v)

i. **Square System: (Plate 3-i)-** In this system planting the plants are planted in straight rows running at right angle. The land between rows being same thus four plants make a square.

ii. **Rectangular System: (Plate 3-ii)-** This system is modification of square system which the plant to plant and row to row distance is not uniform.

iii. **Triangular System: (Plate 3-iii)-** This system is similar to the square system of planting except that in every alternate row the plants are planted in midway of two plants of the previous row. Thus, three plants make a triangle whose only two arms are of equal length.

iv. **Hexagonal System (Plate 3-iv)-** This system is also known as equilateral triangle system of planting. The plants in this system are panted at the corners of the equilateral triangle with one tree in the centre. Thus, six trees make an hexagonal with an additional tree in the centre of the hexagon. This system allows 15 percent more plants than the square system.

v. **Quincunx System: (Plate 3-v)-** This system is essentially the square system except for an additional tree in the centre of each square. Thus the number of trees are nearly double than the square system.

Plate 3: Contour System (Plate 3-vi)

vi. **Contour System (Plate 3-vi)-** In this system the plants are planted by the following the contour of the land and as such suitable for sloppy lands.

Materials and equipments - Peg chain, mallet, khurpi and tall bamboo.

Procedure- For laying out the orchard by any systems of planting, it is necessary that the first row is straight and running parallel to the boundary of the field. The first row is drawn at half the distance within plants. For making first row straight fix bamboos at an interval of 3-4 meters. Stand at one end of the field and see that all the bamboos fall in one straight line, if not, adjust straight and running parallel allowed between rows bamboos by shifting either sides. Continue the process till all the bamboos fall in one straight line. For hexagonal layout, use two chains, after putting the pegs in the first row put chain in two pegs. Hold the chain at desired distance now makes an are from both pegs. Put a peg where both bisect each other. This process is repeated till the entire field is covered.

7

Numerical Problem on Orchard Layout

Problem- Prepare of a layout plan of orchard for 10 ha. area which is situated on the highway. The dimension of the orchard is 3:2. Calculate the areas under each crop which will you grow and also write their suitable variety along with their suitable diagram. Also fulfill the following information about orchard

1. The length and width of orchard
2. Net planting area of the orchard
3. Botanical name of suitable wind break tree
4. Botanical name of a suitable economic hedge plant
5. Suitable fruit crops and their variety for that locality their planting spacing and plant population

Specification about layout of orchard

1. The side ratio of orchard is 3:2
2. Hedge 1m wide around the orchard
3. Wind break 7m wide in north and west direction of sites
4. Main irrigation channel (MIC) 3m wide along with north side
5. Main road 4m wide divide length in two part
6. Sub road 3m wide divide width in two part
7. Sub irrigation channel (SIC) (2) 1.5m wide
8. One fourth (1/4) part of the block 7 or 3 left for farm building (the area of farm building should not be more than 10% of the block)

Calculation

A. Calculation of length and width

Area = 10 ha.

1ha. = $10000m^2$

10 ha. = $10x10000m^2$

$100000m^2$

Suppose to be length and width are in 3:2 ratio and x is common of them

The length is 3x and width is 2x and area is $3x X 2x = 6x^2$

$6x^2 = 100000m^2$

$x^2 = 100000/6 = 16666.67m^2$

$x = \sqrt{16666.67} = 129.09$

Length= 3x = 3x129.09= 387.27m

Width = 2x = 2x129.09= 258.18m

B. Calculation of net planting area

Actual length = Total length- (Hedge + Wind break + Sub irrigation channel+ Main road + Sub irrigation channel + Hedge)

Actual length = 387.27- (1.0+7.0+1.5+4.0+1.5+1.0)

= 387.27-16 = 371.27m

Actual width = Total width- (Hedge + Wind break + Main irrigation channel + Sub road+ Hedge)

Actual width = 258.18 - (1+7+3+3+1)

= 258.18- 15= 243.18m

Area under farm building = Sub plot area/4

= 11285/4 = 2821.25

Net crop area of the third plot lets= 11285-2821.25= 8463.75

Actual cropped area = Actual length x Actual width

371.27 x 243.18 = 90285.43 m^2

Area under layout= Total cropped area – Actual cropped area

= 100000-90285.43 = $9714.56m^2$

Area of each plot lets = Total cropped area/Number of plot lets

$$= 90285.43/8 = 11285.67 m^2$$

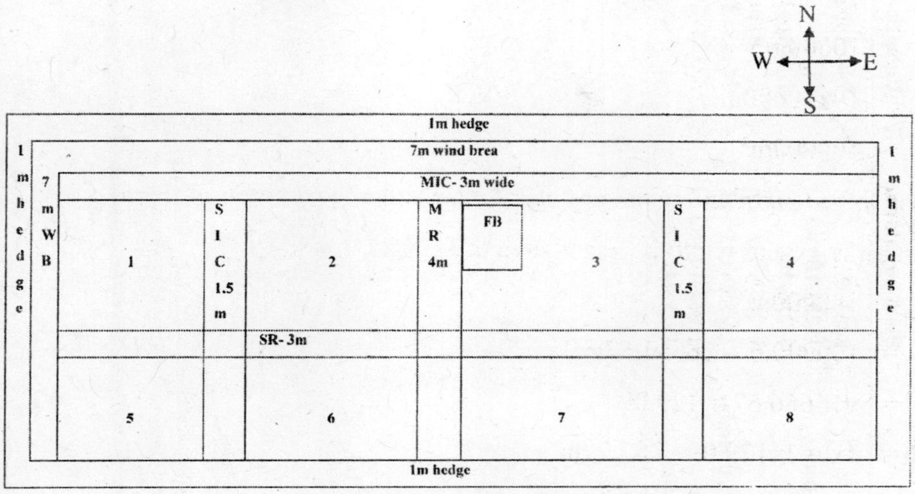

N
W E
S

1m hedge
7m wind brea
MIC- 3m wide
1 m hedge
7 m WB
SIC 1.5 m
M R 4m
FB
SIC 1.5 m
1 m hedge
1 2 3 4
SR- 3m
5 6 7 8
1m hedge

Layout of the orchard

S.N.	Fruit crop	Plot lets	Planting distance	Fruit crops cultivars	Plant population	Remark
1.	Mango	3	3x3m	Amrapali	8463/ 9	940 Approx
		1,2,4	10x10	Langra, Dashehari and Chausa	11285/100 x3	113 each Approx
2.	Jackfruit	5	10x10	Rudrakshi	11285/100	113 Approx
3.	Bael	6	10x10	Narendra Bael	11285/100	113 Approx
4.	Lemon	7/4	5x5	Kagzi lime Eureka and Pant lemon	11285/4x25 =113/3	37+38+38 Approx
5.	Aonla	7x3/4	10x10	NA-7, Chakaiya, Krishna	11285/100 x3	37+38+38 Approx
6.	Guava	8	7x7	Allahabad safeda and L-49	11285/49 x2	115 each Approx

8

Numerical Problem on Mango Orchard

Problem- Prepare a layout plan for establishing on ideal mango orchard near Varanasi city in 2ha area under high density planting. Calculate the number of plants needed for planting and amount of DAP and sulphate of potash heeded or planting. Also mention the name of cultivars for this region under HDP (200:100:200 NPK g/plant)

Area-2ha. = 2 x 10,000m^2 = 20,000 m^2

Plant spacing under HDP = 2.5m x 2.5m

Total no. of plants = 20,000/2.5 x 2.5 = 20,000/6.25

= 3200 plants/2ha

Area = 3 x × 2x = 20, 000m^2

$6x^2 = 20{,}000$

$x^2 = 20{,}000/6 = 3333\text{m}^2$

$x^2 = \sqrt{13333} = 57.4$ m

Length-3x = 3 x 57.7 = 173.1m

Width-2x = 2 x 57.7 = 115.4m

Total area - 173.1 x 115.4 = 19975.7m^2

Area of main road = 2.5m x 115.4m = 288.5m^2

Main irrigation channel = 1.5m x 173.1m = 259.65m^2

Bunds area = 0.30 x 173.1 x 4 = 207.72m^2

= 0.30m x 115.4 x 4 = 138.48m2

Store room area = 10m x 10m = 100m^2

Well area = 10m x 10m = 100m^2

Total = 1094.35 m^2

So actual area = 19975.7-1094.35m^2 = 18881m^2

So, total plants = 18881/2.5 x 2.5 = 3020 plants /2ha

Total N to supply for 3020 plants = 200g x 3020 = 604 kg

Total P_2O_5 to supply for 3020 plants = 100g x 3020 = 302 kg

Total K_2O to supply for 3020 plants = 200g x 3020 = 604 kg

Amount of DAP to apply N/18; P/46 (46 kg P needed in 100 kg DAP)

= 100/46 x 3020 = 15000/23

= 656.5kg = 657kg DAP

100 kg DAP contain 18 kg N

657kg DAP found N 18/100 x 657 = 118.3 = 118kg N

So, remaining N_2 to supply = 604-118 = 486 kg N

(a) Amount of Urea to supply = 100/46 x 486 = 2.2 x 486 = 1069 = 1069 kg Urea

(b) Amount of Sulphate of potash to supply = 100 x 604/60 = 1000 kg

9

Planting Board

Purpose: It is important that the plants occupy the same position where the pegs were placed in the field. At the time of planting the tree the peg is removed first followed by pit making to receive the plant. The plant may be with the ball of earth or without the ball of earth. While planting the tree in the pit there is every possibility that plant may not occupy the same position where the pegs were originally placed. In order to overcome this problem planting board is used. Use of a planting board ensures the planting of the plant in the correct position.

Materials and equipments: Spade, khurpi, planting board, wooden pegs and chain.

Procedure: Having assumed that the planting plan has been selected, land is prepared and been fixed at the correct positions. For planting trees, remove the peg and dig a pit for receiving the ball of the earth, while planting the plant in the pit the trunk of the plant should occupy the position of peg. Plant planted with the help of planting board (Plate 4) ensures panting of the plant in the correct position. Place the planting board n such a manner that the centre notch of the board is in the position of marking peg (Plate 4-i-iv). Drive in pegs at each end of the board. Now remove the planting board as well as the marking peg. Dig a hole with the help of khurpi large enough to receive the ball of earth, where the marking peg stood. Put the tree with ball of earth in the pit. Bring back the planting board in such a manner that the two pegs (guide pegs) at both ends position of the tree. The tree should occupy the same position which was occupied by the marking peg. After adjusting the position which was occupied by the marking peg. After adjusting the position of the tree, adjust the depth of the plant. The plant should be planted at the same depth as they stood in the help of khurpi handle. The pit should be filled up to the ground level, make a small basin around the tree and irrigate the plant. All the new growth as well as the surplus growth at the time of planting is to be removed. Planting should be done either in the evenings or on a cloudy day.

Planting Board & Its Use

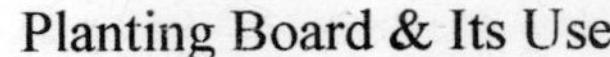

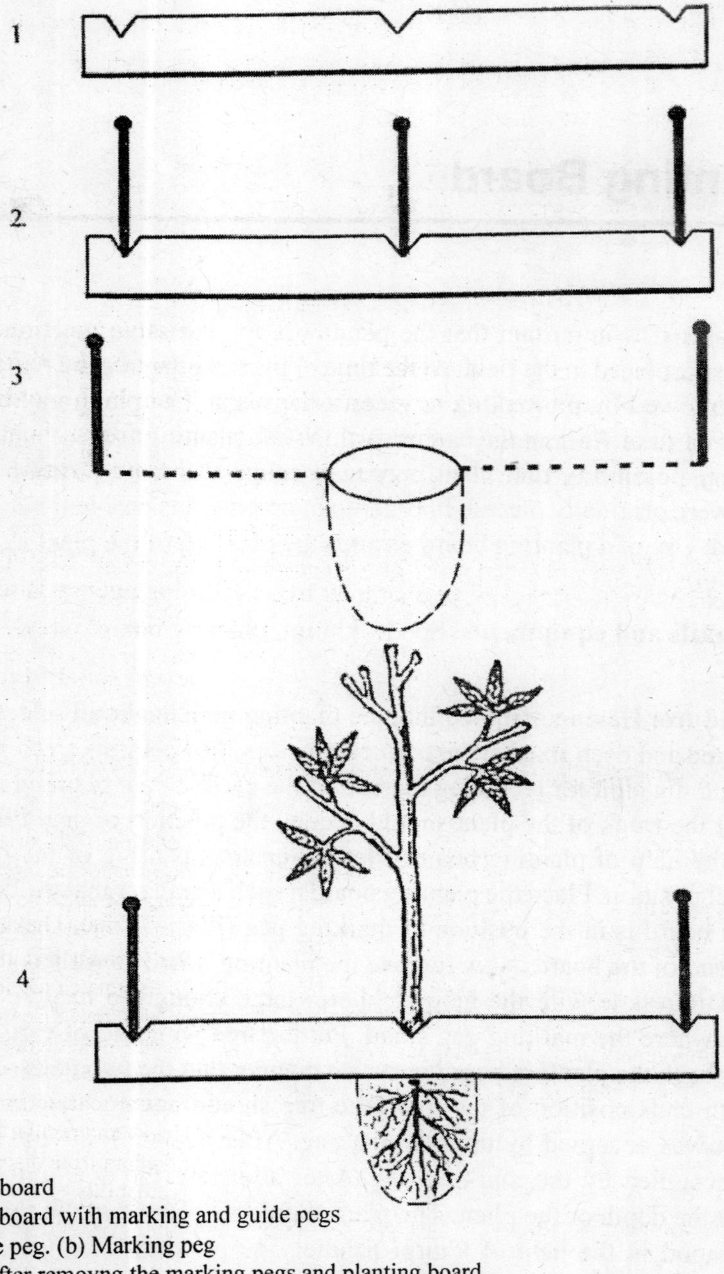

1. Planting board
2. Planting board with marking and guide pegs
   (a) Guide peg. (b) Marking peg
3. Pit dug after removng the marking pegs and planting board.
4. Planted plant adjusted by bringing back the planting board.

Plate 4: Planting Board and Its Use

10

Mother Plant and Bud Wood

Purpose: In propagation of fruit plants by any vegetative methods such as cuttings, budding and in to the selection of 'mother plants" and "bud-wood". Use of bud-wood or any other propagation material from high yielding, true-to-type, disease free trees (mother plant) will ensure uniform behaviour of trees in the orchard. If possible, propagation material from a single source is used.

Bud-wood used for budding purpose from high yielding, true-to-type and disease free trees certified for the above character by a certifying agency is referred as 'certified bud'

Procedure: By adopting the flow sheet given below, clean scion from healthy, productive plants, can be selected for vegetative propagation.

Table- Flow sheet of selection of mother plants and bud-wood

1st Step- Orchard map-Consideration	1. Location of the orchard 2. Location of tree in the orchard 3. Details about scion 4. History of the orchard 5. Soil, climate and topography
2nd Step- Orchard survey -Purpose	1. To eliminate un-uniform trees 2. To eliminate diseased trees
3rd Step- Preliminary inspection of the individual trees- Consideration	1. Healthy and vigorous trees 2. True to type 3. External characters of fruit 4. Heavy and regular bearing 5. Freedom from diseases (virus) 6. Availability of bud wood
4th Step- Selection of wood from mother plants-Through	1. Study of records for yield and quality 2. Elimination of pants (unproductive or irregular bearer)

Contd.

5th Step-Selection of mother plants- Through	1. Study of records for yield, fruit maturity and uniform size of fruit
6th Step-Selection of bud wood- Consideration	1. Plumpness of wood 2. Freedom from thorn 3. Avoid flower-bud (Leaf buds are small and sharp pointed) 4. Round and lead pencil thickness.

11

Propagation by Cutting

Purpose- Multiplication of plants by cutting includes stem, root and leaf cuttings. The stem cuttings are of four different types i.e. hard, semi hard, soft wood and herbaceous wood cutting consist in producing new individuals after the piece of stem has been detached from the mother plant and planted in conditions suitable for regeneration to the missing organs. Stem cutting have to develop a new root system as they already have buds to give rise to shoots. Many cells of the cutting are capable of becoming meristematic and thus are able to produce new roots. The success in the stem cutting multiplication depends upon factors such as condition of the mother plants, part of the tree from where the cuttings are made, time of the year, care while planting and after care. Cuttings can be made during monsoon (July-Sept.) and spring (Feb. -March) season when active cell division is taking place.

CLASSIFICATION OF CUTTINGS

A. Stem cuttings

1. Hard wood
2. Semi-hard wood
3. Soft wood
4. Herbaceous cuttings

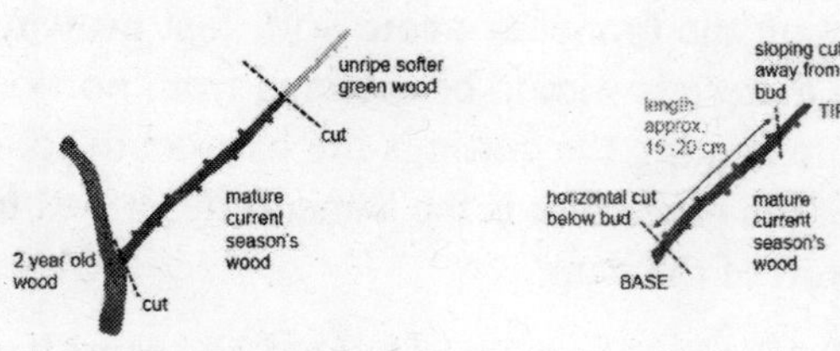

Plate 5

B. Root cuttings

C. Leaf cuttings.

A. HARD-WOOD CUTTINGS

Materials and equipments- Secateour, knife, khurpi, rooting media and pots or flats.

Procedure- Select branches of current year or of post season growth, about lead pencil thickness from healthy, vigorous and young plants. Use these branches for making cuttings of 20 to 25 cm in length. The basal cut is given about 0.3 cm below a bud, and the upper cut about 1.0 cm above the bud. Remove the leaves from the basal two-third parts of the cuttings. Make holes in the rooting media at 2.5 cm apart and 5.0 cm between rows. Plant cuttings in the holes. Bury the two-third basal part of the cuttings in the holes. Press, soil around cutting firmly. If the rooting media is dry sprinkle water before making the holes. Sprinkle water on cuttings. Keep the pots in a cool, moist place for rooting. Sprinkle water as and when necessary. Record the data as outlined below.

B. SEMI-HARD WOOD CUTTINGS

Procedure- Select healthy, disease-free vigorous plants as parent trees. Severe semi-mature branches from the selected tree. Make cuttings of 10 to 15 cm in length. Remove the leaves only from the two-third basal part of the cuttings. As wood in immature cuttings have poor storage of carbohydrates and also of auxins which are essential for rooting. While making the cuttings the basal cut should be given about 0.3 cm below the node and the upper cut about 1.0 cm above the node. Rest of the procedure is the same as described for hard-wood cutting including the recording of data.

C. SOFT-WOOD (GREEN WOOD) CUTTINGS

Procedure- Remove immature branches from the selected plants which are of 4-6 months old. Do not remove the leaves except for the part to be burried inside the rooting media. In selecting the branches avoid soft, fast growing, weak and interior shoots. Planting materials should be selected from portions of plant growing in full light. While making the cuttings the basal cut should be given just below a node. Rest of the procedure is the same as described for hard wood cuttings, including record of the data.

Soft-wood cuttings root easily and quickly as compared to hard and semi-hard wood cuttings. This is possible however, only after proper care and preventing the desication of the tissues. Soft-wood cutting should be kept in green or polythene houses or in a moist chamber where high humidity is maintained, which keeps the tissues in turgid condition.

A cheap but very effective humid environment can be created easily by covering a pot with a polythene bag (150-20 guage). The pot should be fully covered by the bag. In order to hold the bag in its position it needs the support of a wire loop. The ends of the wire needs to be stuck in the pot. This device proves very effective for propagation soft-wood and herbaceous cuttings.

TREATMENT OF THE CUTTING WITH GROWTH REGULATORS

Cuttings when treated with root-promoting chemicals to hasten root initiation and increase the number of roots. The growth regulators which are extensively used for treating cuttings along with their dilution and methods of application are given in table 14-A and 14-B.

Application technique of growth regulators and their duration

1. **Soak treatment-**

   a. Prolong Dip Method (Dilute solution) 18-24 hours.

   b. Quick Dip Method (Concentrated solution) 5-10 seconds.

2. **Talc or dust treatment**- Moistened basal end of cuttings are dipped in the dust containing growth regulator. The concentration of active ingredient varies from 500-1000 ppm.

Selection of growth regulators- The following growth regulators are suggested for encouraging rooting in stem cuttings (Table14-A).

Table 14 (A)

S.No.	Name of growth regulators with abbreviations	Suggested concentration (ppm)		Remark
		Quick Dip	Prolong Dip	
1.	Indole acetic acid (IAA)	500-1500	100-500	Effective over a large number of species. Roots are thick
2.	Indole butryic acid (IBA)	500-1500	100-500	Excellent for rooting as effective over a large number of species. Not harmful even on higher concentration. Root are fibrous.Moves slowly in the plant and is destroyed slowly by auxin destroying enzyme

Contd.

3.	Naphthalene acetic acid (NAA)	500-1500	100-500	Good for rooting certain king of plants. Toxic at higher rate
4.	2,4-dichlorophenoxy acetic acid (2,4-D)	10-100	5-20	Good for rooting at very low concentrations but inhibits the shoot growth at high concentration. Produce bushy, stunted and thick roots.
5.	2,4-5 Trichlophenoxy acetic acid (2, 4, 5-T)	10-100	5-20	Good for rooting at very low concentrations but inhibits the shoot growth at high concentration. Produce bushy, stunted and thick roots.

TREATING CUTTINGS WITH GROWTH REGULATOR SOLUTION

Procedure- The prepared cuttings should be treated with the growth regulator solution as soon as possible. Put the growth regulator solution in beakers. Label the beakers properly with glass marking pencil. Make bundles of ten or more cuttings for each concentration of each growth regulator solution. Put the basal ends of cuttings in the solution. At least 2-3 cm of cuttings should be inside the solution. Leave the cuttings in the solution to the desired duration. In one beaker, take distilled water and keep same number of cuttings and for same duration as allowed for the growth regulator solution. After the required period, remove the cuttings from the solution and wash them with plain water. After this, plant the cuttings as described for hard-wood cuttings. Record the data as outlined in blank 14-2, and 14-3. Note the difference in rooting between the treated and untreated cuttings. The better rooting is due to the growth regulators

Conclusion

Treated cutting roots earlier than the untreated ones. However, certain species do not root even after the treatment. Treated cutting produce more number of roots. Semi-hard and soft wood cutting perform better than the hard wood culture. Maximum success is obtained during period of active cell division.

Table 14 B: Calculation of strength of growth regulators

Percent solution (%)	Milligram 1 litre (mg/l)	Parts per million (ppm)	Gram/litre* (g/l)
100	1,000,000	1,000,000	100
10	100,000	100,000	10
1	10,000	10,000	1
0.1	1,000	1,000	0.1
0.01	100	100	0.01
0.001	10	10	0.001
0.0001	1	1	0.0001

***(1) 1.0 percent = 10,000 ppm, (ii) ppm = percentage x 10,000 (iii) percentage**

$$= \frac{\text{ppm}}{10,000}$$

12

Propagation by Layering

Purpose: Layering includes several forms of ground and aerial layering (goottie). When rooting is encouraged on the aerial part of a plant after wounding it is known as air layering or goottie or morcottage. When branches running parallel to ground are utilized, then the method is known as ground layering. Layering is the development of root on a stem while it is still attached to the mother plant. The root formation on the stem is stimulated by removing a ring of bark or by making a notch which cause an interruption in the downward movement of carbohydrates and auxins from the leaves and growing shoot tips. The materials accumulate near the points of treatment and helps in rooting. Root formation depends upon continuous moisture supply, good aeration and moderate temperature around the rooting zone. For the same reason the best result are obtained by these methods during monsoon month (July-Sept.) and spring (Feb. -March) from branches 1 to 2 years of age.

CLASSIFICATION OF LAYERING

A. Ground Layering

1. Tip layering,
2. Simple layering,
3. Trench layering,
4. Mound or Stool layering,
5. Compound or Serpentine layering

B. Air layering (Goottie or mercottage)

A. GROUND LAYERING

1. Simple Ground Layering

Materials and equipments-Grafting knife, khurpi, and secateour.

Procedure- Bend the cane tip down to ground level for multiplication. Remove a ring of bark of make a notch only at 20 to 25 cm away from the growing tip. Remove the ground soil up to 7.5 to 10 cm depth. Bury the injured cane portion in the soil. Put the soil again on the injured portion. If necessary put a piece of stone to hold the cane in its position. Keep the soil wet where cane is buried for developing the roots. The roots will develop within 4-6 weeks, remove the rooted cane and plant it in a cool shady place.

2. Compound or Serpentine Layering

Procedure- Bend the cane and bring down to the ground level (Plate 6). Make 3 to 4 notches or give wound at each node of the cane at a distance of 10 to 20 cm apart. Bury the wounded nodes in the hole prepared for the purpose. This way all the wounded nodes are buried. If necessary place a piece of stone to hold the cane in its respective positions. Leave 20 to 25 cm of tip above the ground. Detach the new plants after 45-60 days, after separating each rooted portion plant them in nursery.

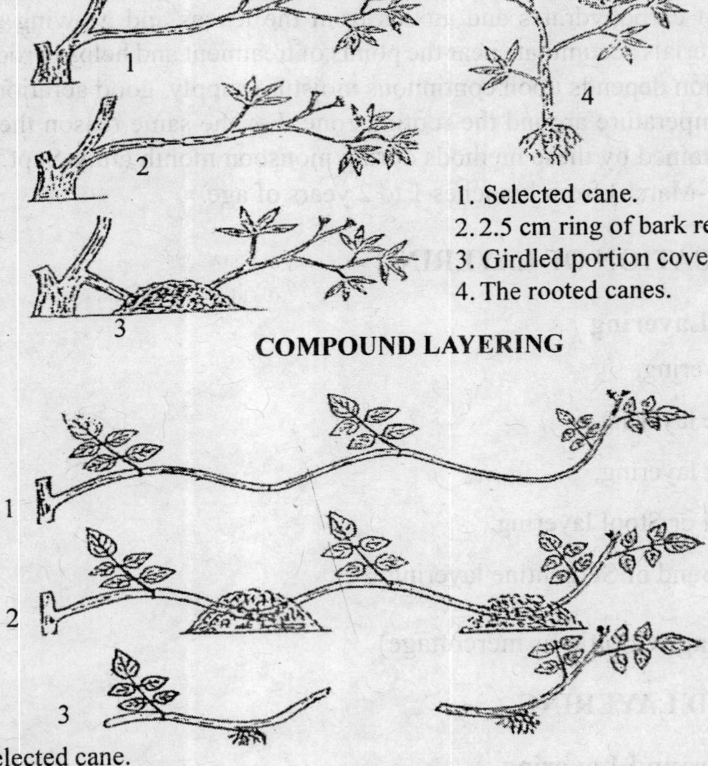

1. Selected cane.
2. 2.5 cm ring of bark removed at two places and covered with soil.
3. The rooted canes.

Plate 6:

B. AIR LAYERING

Materials and equipments- Grafting knife, moss grass, polythene film (200 to 300 guage) and string..

Procedure- Select branches of one or two years of age and about a lead pencil thickness. Wound the branch by girdling ust below a node about 25-35 cm away from the growing tip. Remove the bark completely from the girdled area (about 2.5 to 3.5 cm). Place handful of moist not wet moss grass around the wounded barkless area. (The moss grass can be substituted with a mixture of soil, sand and leaf mould in ratio of 2:1:2 respectively). Wrap carefully with a piece of clear polythene film 20-25 cm in size. To cover the moss grass completely. Gunny bag pieces of 20-25 cm can substitute for polythene film as a wrapping material. Tie both the ends tightly. In case wrapping material is gunny bag frequent watering in essential. But in case of polythene film no further watering is needed. Polythene film permits the exchange of gases (carbon dioxide and oxygen) and low transmission of water vapour.

Severe the air layers after 45-60 days when there are plenty of roots, which can be easily seen through the film or emerging through gunny bag. Plant the air layers in a cool shady place.

STEPS USED IN AIR LAYERING

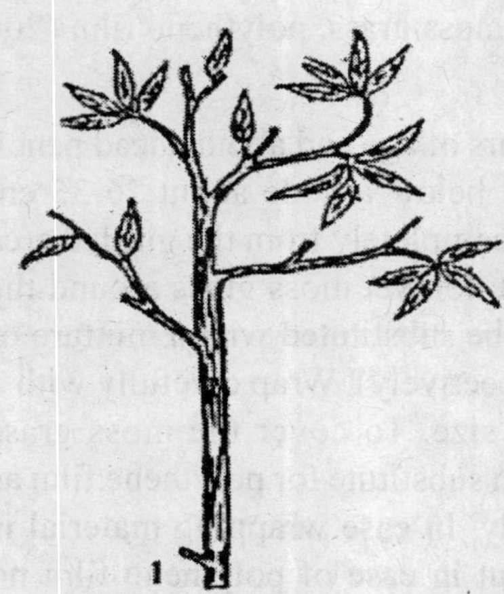

1. Selected branch.

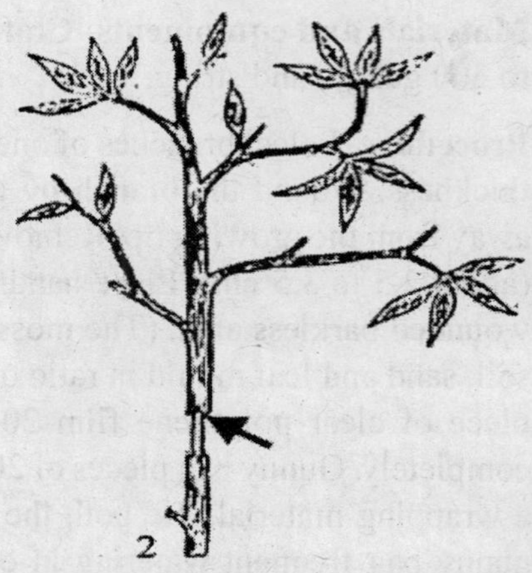

2. 2.5 cm area of bark removed (Arrow indicates the place of growth regulator application).

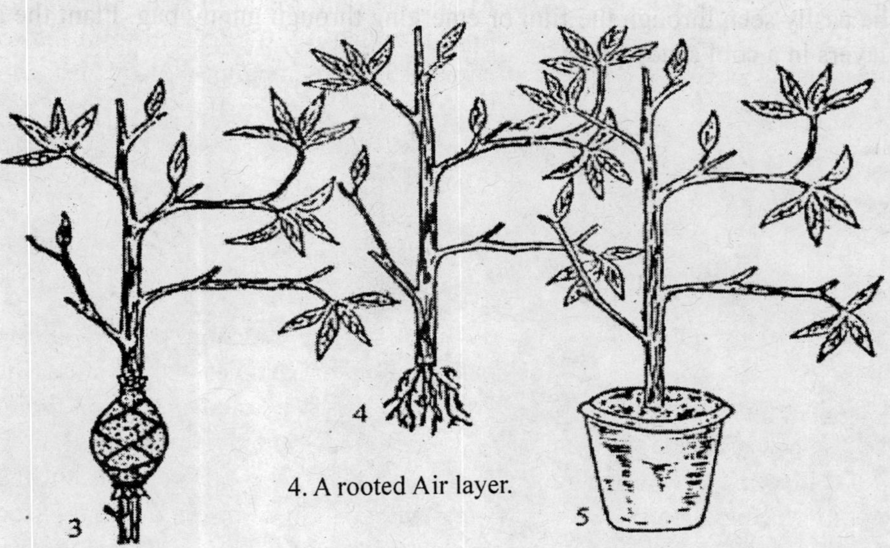

3. Ringed section covered with moss grass and wrapped with a sheet of polythene film, ends tied.

4. A rooted Air layer.

5. Air layer potted.

Plate 7: Steps Used in Air Layering

13

Propagation by Budding

i. SHIELD OR 'T BUDDING

Materials and equipments- Budding knife, tying material and petridis.

Procedure

Operations on the stock- Select healthy, uniform stock plants of about a lead pencil thickeness and about one to one and half years of age. Remove the thorns (if any) and side branches up to a height of 20 to 25 cm from the soil surface. Select a spot at height of about 15 to 20 cm from the soil surface in between two internodes, which is smooth, other than the southern aspect. With the help of a sharp knife make a horizontal cut of about 1.25 to 1.80 cm in length, deep enough to cut bark only. Starting from the centre of the horizontal cut make a vertical cut, deep enough to cut the bark only of about 2.5 to 3.5 cm long making the shape of english letter 'T'. Open the flaps with the help of knife blade or the back of the budding knife.

Operations on the scion and tying- Slice out the bud from the bud stick of about 2.5 to 3.5 cm long with the consideration that bud lies in the centre. Slice deep enough to include a thin layer of wood also. Take out the wood though this is not necessary. Now this shield-shaped but is ready for inserting in the 'T' shaped cut given on the stock. Several buds may be prepared but they should be kept in petri-dish containing water and to be inserted into the stock later. Hold the bud in such a manner that the eye of the bud is facing upward. Push down the bud into the 'T' shaped cut made on the stock. Continue pushing of the bud till the entire bud is covered by vertical cut on the stock and the eye of the bud lies almost in the centre. Tie the bud firmly but not so firmly that will impede the flow of sap. Tying may be started from any end. While tying care is to be exercised that a space of about 0.5 cm above the leaf petiole is left for the growth of bud. For tying any of the materials can be used, e g., sutli, banana fibre and polythene strips. The polythene strips of 200 or 300 guage are better tying material as the buds sprout between earlier and the percentage of success is also higher. Care should be taken that very little time is elapsed between opening of bark of the stock and inserting of bud. The various steps involved in the operation are illustrated in Plate 8.

In case of inverted 'T' budding, transverse cut is given at the bottom of the vertical cut. Rest of the procedure is the same as described above, except that the shield piece containing bud is inserted into the vertical cut from the lower part of the cut. The bud is pushed upward till the entire cut is covered.

Steps Used in 'T' Budding

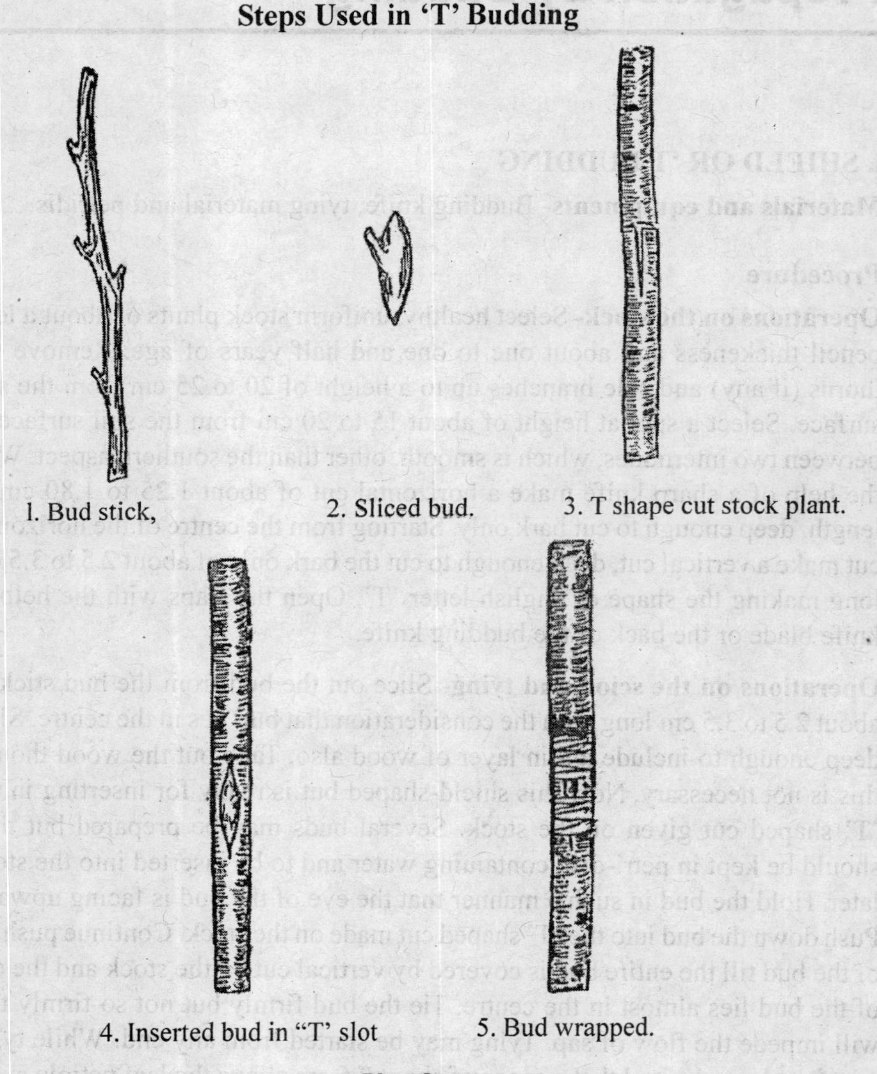

1. Bud stick, 2. Sliced bud. 3. T shape cut stock plant.

4. Inserted bud in "T' slot 5. Bud wrapped.

Plate 8:

(ii) PROPAGATION BY PATCH BUDDING

Materials and equipment- Patch budding knife, polythene film and petridis.

Procedure

Operations on the stock- Select rootstock one to one and half years old about a lead pencil thickness, about 15 to 20 cm above the surface soil, make a horizontal cut of 1.25 to1.80 cm length on the rootstock deep enough to cut the bark only. About 2.5 to 3.5 cm apart from the horizontal cuts make another cut of the same size as the previous one. Two horizontal cuts at 2.5 to 3.5 cm apart can be easily made with the help of a patch budding knife. Join these parallel cuts by two vertical cuts connecting the ends of the horizontal cuts. Thus a rectangular patch of the bark made above is removed from the stock plant.

Operations on the scion and tying- From the selected scion branch remove a rectangular piece, of the bark containing one eye on the same size as given on the stock with the consideration that the bud lies in the center of the bark patch. Use of patch budding knife will ensure the removal of patch of bark (containing eye) of the same size as that of stock. Place the scion patch of bark on the stock from where a similar patch of bark was removed. Ensure that all the four sides of scion patch are in close contact. Tie the bud with the polythene film leaving room 0.62 cm for the bud to resume its growth. The various steps involved in the operation of patch budding are illustrated in Plate 9

Step Used in Patch Budding

I . Bud stick.

2. Patch of bark removed with bud.

3. Patch of bark to be removed.

4. Patch of bark removed.

5 - Patch fitted on the stock .

6. Patch bud wrapped.

Plate 9:

(iii) FORKERT METHOD OF BUDDING

Materials and equipments- Budding and tying material (Polythene film).

Procedure

Operations on the stock- Select stock plant one to one and half years of age aoout 1.2 to 1.8 cm in diameter. Select a smooth surface on the selected root-stock at a height of 15 to 20 cm above the soil surface. With the help of a sharp knife make transverse cut of 1.2 to 1.8 cm in length, deep enough to cut the bark only. From the ends of the transverse cut make two vertical cuts downward of about 3.5 to 5.0 cm in length. Pull down the rectangular panel (flap) of the bark carefully.

Operations on scion and tying- Select the scion braches of the same thickness as that of the stock. Remove the leaves close to the bud. From the prepared scion branch a rectangular piece of bark containing only one bud (eye) of 1.25 to 1.8 cm in diameter and 3.5 to 5.0 cm in length is removed with the help of sharp knife. If any wood remains adhering to the bark it should be removed. Place the bud on the stock in such a manner that the patch of bark containing the bud rests in the angle made by the stock flap and wood. After placing the bud, the bark flap is pulled up to its original position covering the bud entirely. Tie the bud with wrapping material, e.g. wax-cloth, rubber band of polythene tape with care that the bud is not tied tightly. Untie the bandage after fifteen or twenty days and examine the bud. If the bud is green, then cut the stock flap just below the bud and wrap it again but less tight. The various steps involved in the operations are illustrated in Plate 10.

Step Used in Forkert Budding

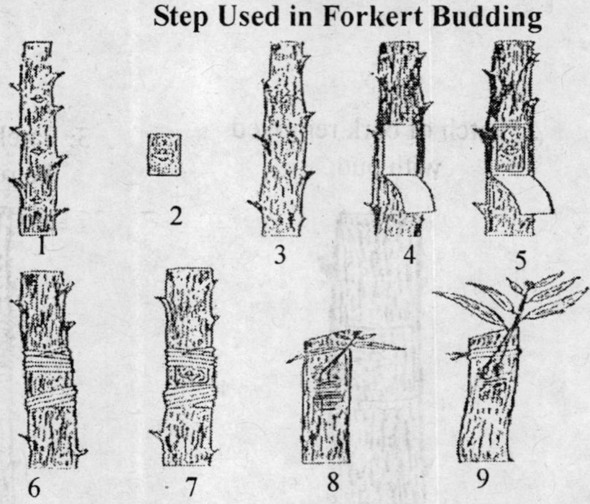

1. Prepared scion branch. **2.** Rectangular bark: with one bud removed from bud stick. **3.** Prepared root stock. - **4.** Operated stock. **5.** Bud placed on the stock. **6.** Bud wrapped. **7.** Flap of the stock removed below the bud. **8.** Stock headed back. **9.** Trained scion shoot.

Plate 10:

(iv) PROPAGATION BY RING BUDDING

Materials and equipment- Budding knife and wraping material.

Procedure

Operation on the stock- Select healthy stock plants of one to one and half years of age of about a lead pencil thickness. Head back the plants at a height of 1.0 to 1.5 cm from soil surface. Remove a ring of bark 4 to 6in length from the top of the stock (see plate 11).

Operation on the scion and tying- Selected scion branches of the same thickness as of stock. Remove the leaves, keeping the petiole intact. From the basal end of the scion branch, remove a ring of bark containing an eye. The ring of bark to be removed should be of same size as that of stock. For removing the ring of bark from the scion and stock, girdled portion is first twisted followed by pulling up the snug.

Now, slip the scion ring down over the barkless stock, till it reaches the bark. If the ring does not fit tightly, more bark from the stock is removed and ring lowered. This process continues till the scion ring fits securely on stock plant. Tie the bud tightly leaving enough space for sprouting of bud. The various steps involved in the operation of ring budding are illustrated in Plate 11.

Steps Used in Ring Budding

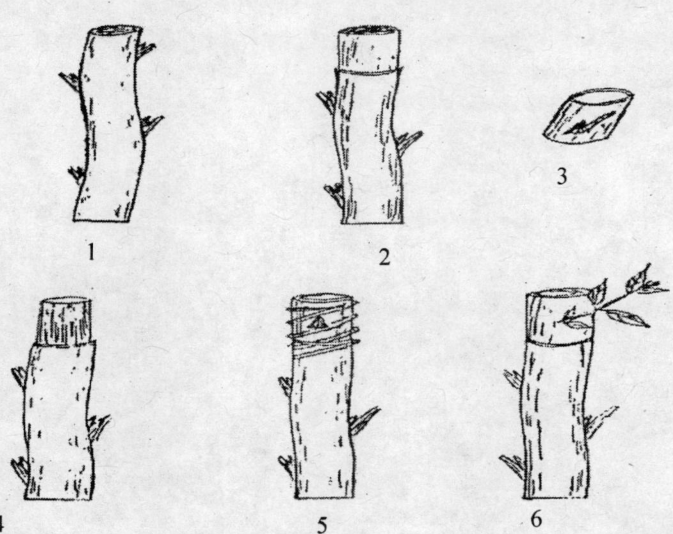

1. Prepared scion branch. **2.** Ring of bark: containing one eye removed from scion branch. **3.** Ring of bark containing an eye., **4.** Ring of bark removed from stock. **5.** Scion ring fitted on stock. **6.** Sprouted scion.

Plate 11:

(v) PROPAGATION BY RING BUDDING

Materials and equipment- Budding knife and wraping material.

Procedure

Operation on the stock- Select healthy stock plants of one to one and half years of age of about a lead pencil thickness. Head back the plants at a height of 1.0 to 1.5 cm from soil surface. Remove a ring of bark 4 to 6 cm length from the top of the stock (see plate 11).

Operation on the scion and tying- Selected scion branches of the same thickness as of stock. Remove its leaves keeping the petiole intact. From the basal end of the scion branch, remove a ring of bark containing an eye. The ring of bark to be removed should be of same size as that of stock. For removing the ring of bark from the scion and stock, girdled portion is first twisted followed by pulling up the same.

Now, slip the scion ring down over the barkless stock, till it reaches the bark. If the ring does not fit tightly, more bark from the stock is removed and ring lowered. This process continues till the scion ring fits securely on stock plant. Tie the bud tightly leaving enough space for sprouting of bud. The various steps involved in the operation of ring budding are illustrated in Plate 11.

Steps Used in Ring Budding

1. Prepared scion branch. 2. Ring of bark containing one eye removed from scion branch. 3. Ring of bark containing an eye. 4. Ring of bark removed from stock. 5. Scion ring tied on stock. 6. Sprouted scion.

Plate 11:

14

Propagation by Grafting

Purpose- The term 'Grafting' includes all forms of grafting and budding. It is the art of joining the part of together in such a manner that they will unite and continue their growth as one plant. A grafted plant consist of two parts a rootstock, also known as understock or stock, and a scion. The part of the graft combination which is to become the upper portion or top and performs the function of flowers production and fruits is known as scion and the part which is lower portion or root which performs the function of fixing itself in the soil, absorbs moisture and nutrient from the soil is known as rootstock.

When the scion part is a small piece of bark containing only one bud the operation is termed as budding, but when the scion part is a branch containing more than one bud, the operation is termed as grafting.

The budding and grafting is possible between varieties in a species or between the species of a genus and some times between different genera of the family. Better results are expected when the rootstock and scion are of equal thickness and are of one to two years of age. The best time for grafting is monsoon (July to August) and for budding is the spring (February to March) for evergreen plants whereas, for deciduous plants the period when they are dormant is the best time for grafting.

The budding and grafting makes possible (1) to multiply varieties that cannot be multiplied by cutting, layering or division, (2) to change of the top of tree usually of an desirable variety, (3) to obtain the benefits of rootstocks or intermediate stock, (4) to obtained higher yield, (5) to overcome the problem of disease and pest, and (6) to overcome the problem of soil.

ESSENTIAL OF GRAFTING AND BUDDING

The success in grafting or budding depends upon the observance of the following.

1. The scion and stock must be congenial or capable of growth when properly united suitable conditions.

2. The operation must be done at the proper season of the year.
3. The growing tissue (cambium) of the scion should be in close contact with the growing tissue (cambium) of the stock.
4. To prevent drying out, all wounded surfaces must be properly protected.

CLASSIFICATION OF GRAFTAGE

1. **Attach grafting-** Approach grafting
2. **Detached grafting-** Whip or tongue grafting, Cleft grafting, Side grafting, Veneer grafting, Splice grafting

(i) APPROACH GRAFTING

Material and equipment- Budding knife, secateurs and tying material.

Procedure- Selected stock plant of one and half year of age. Place the stock plant along the side of the mother plant (scion plant). Selected a branch from the mother plant of the same thickness as that of stock. Place the scion branch running parallel to the stock plant. Bring the stock and scion together. Mark out an area of 3.5 to 5.0 cm in length, where stock and scion pieces meet together easily. Slice out a thin layer of bark along with wood from the marked area on the stock. Give similar cut of same size on the scion branch as on stock from the marked area. The depth of the cut given on stock and scion should not be deeper than the one-third of the total thickness of the piece. If stock and scion are the unequal diameters then the thinner one should be sliced deeper, so that the cambium layers meet easily

Bring together two cut surfaces as soon as possible. Tie the cut surfaces together in such a manner that no space is left in between them. The tying can be accomplished either by sutli or was cloth or polythene strips. After 6 to 8 weeks, when the union has completed, the stock is cut above the bandage and the scion below the bandage. Keep the newly grafted plant in a cool and shady place. The various steps involved in the operation are illustrated in plate 12.

Steps Used in Spliced Approach Grafting

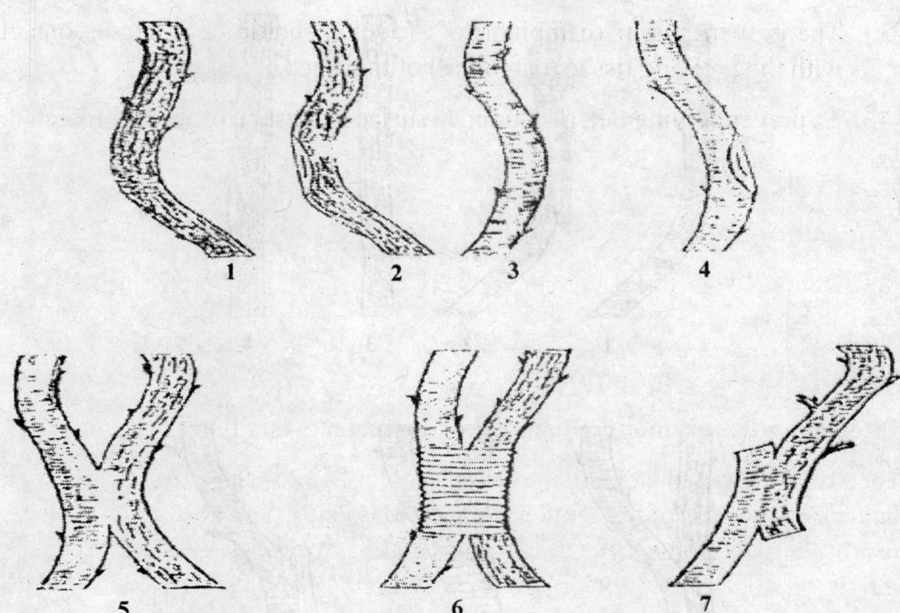

1. Scion. **2.** Scion with slice of bark &. wood removed. **3.** Stock- **4.** Stock with slice of bark and wood removed. **5.** Stock and sc ion united. **6.** Stock and scion tied. **7.** Stock top removed and scion detatched from the mother plant.

Plate 12:

(ii) TONGUE GRAFTING

This method is differed from the simple approach grafting in the manner that a tongue is made on both stock and scion for better cambial contact. Prepare the stock and scion exactly in the same manner as described above. Starting from the middle of the sliced area of the stock give slopping cut into the wood above 1.25 cm deep in the scion pointing downward. Place the two cut surface in such a manner that tongue of each other are inserted in the slit of each other. The trying is as usual. The various steps involved in the operation are illustrated in Plate 13.

Steps used in Tongue Approach Grafting

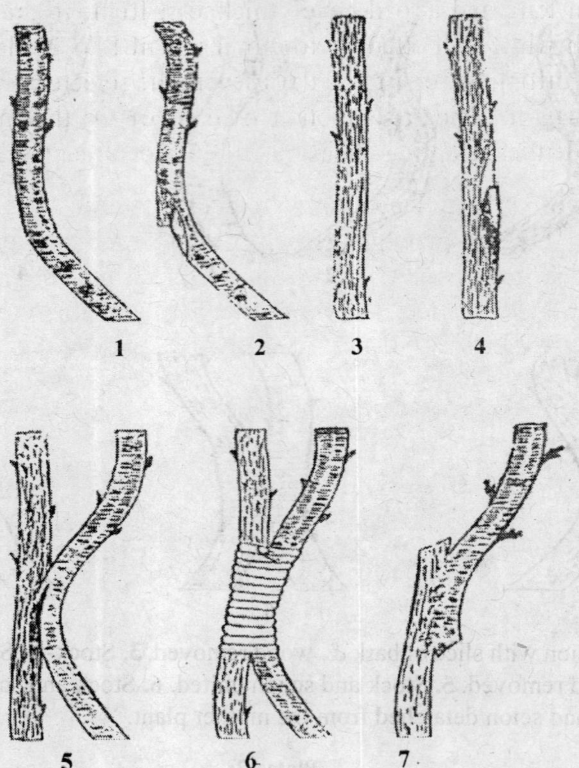

1. Scion.
2. Scion with slice of bark and wood removed and tongue made.
3. Stock.
4. Stock with slice of bark removed and tongue made.
5. Stock and scion united.
6. Stock and scion tied.
7. Stock top removed and scion detatched from the mother plant.

Plate 13:

(iii) PROPAGATION BY VENEER GRAFTING:

Materials and equipment: Same as propagation by budding.

Procedure

Operations on the stock- Select stock plant of 10 to 12 months old of 1 to 2 cm in diameter. Make a shallow cut of about 2.5 to 3.0 cm on one side of the seedling plant at a height of about 10 to 15 cm from the soll surface with the help of a sharp knife. Remove the bark along with wood thus making an oblique cut. The cut should not be deeper than one-third thickness of the stock at the lower and end of the slanting cut.

Operations on the scion and tying- Select scion branch of the desired variety of 10 to 15 cm long and a lead pencil thickness. Remove the leaves of the selected branch while still attached to mother plant 8 to 10 days before the actual date of grafting. After 8 to 10 days severe the selected branch from the mother plant (day on which operation is to be done). On the basal end of the scion give an identical slanting cut as that of the stock plant.

Place the cut parts of the scion along with that of stock in such a manner that the cambiums of both are matching closely. Tie the union portion tightly either with wax cloth or polythene of 200 gauges of 1.25 to 1.80 cm wide strips leaving the terminal ends free. The various steps involved in the operation are illustrated in plate 14.

Steps Involved in Veneer Grafting

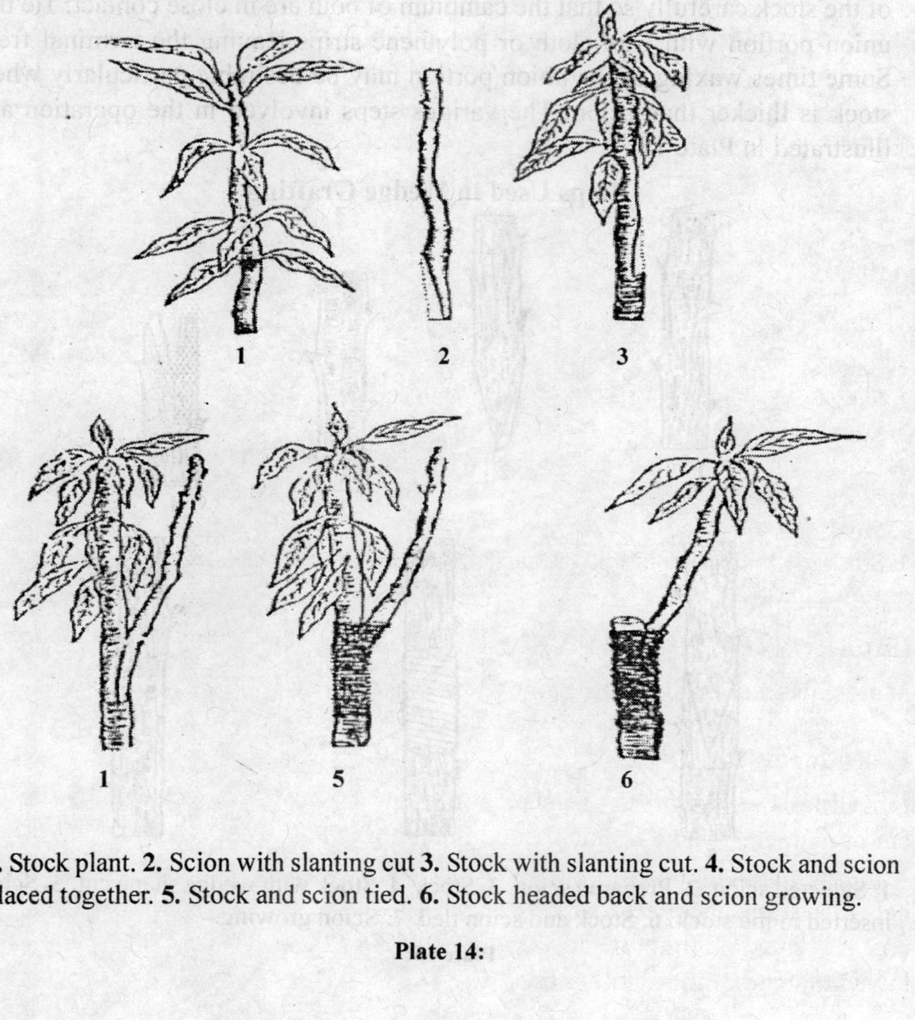

1 2 3

1 5 6

1. Stock plant. **2.** Scion with slanting cut **3.** Stock with slanting cut. **4.** Stock and scion placed together. **5.** Stock and scion tied. **6.** Stock headed back and scion growing.

Plate 14:

(iv) PROPAGATION BY CLEFT GRAFTING

Materials and equipment- Same as propagation by budding.

Procedure

Operations on the stock- Select stock plant of 12 to 14 months old of 1.5 to 2.0 cm in diameter. Head back the stock plant at a height of 20 to 25 cm from the surface of soil with the help of secateour. Smoothen the cut surface of the stock with the help of a sharp knife. In the centre of the be headed stock made a 'V' shaped cut with the help of a sharp knife.

Operations on the scion and tying: Select scion branches of the desired variety of about 10 to 15 cm long, having at least 3 to 4 buds. Prepare the scion by removing the leaves. Make a wedge-shaped cut or the lower end of the scion of similar size as that of stock. Place the wedge of the scion in the 'V' shaped cut of the stock carefully so that the cambium of both are in close contact. Tie the union portion with wax cloth or polythene strips leaving the terminal free. Some times waxing of the union portion may be desirable particularly when stock is thicker than scion. The various steps involved in the operation are illustrated in Plate 15.

Steps Used in Wedge Grafting

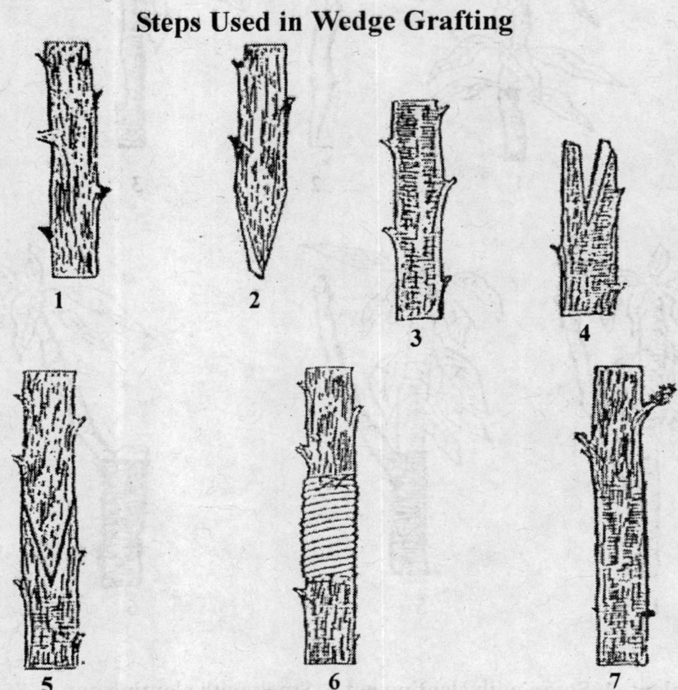

1. Selected scion. **2.** Prepared scion. **3.** Stock. **4.** Stock with wedge shape cut. **5.** Scion inserted in the stock. **6.** Stock and scion tied. 7. Scion growing.

Plate 15:

(v) PROPAGATION BY EPICOTYLE GRAFTING

Materials and equipment- Grafting knife, secateurs and tying material.

Procedure- Germinating seeds about 4 to 8 days old are used as root stock, when it is bronze in colour. Stock in removed at 3cm from the stone or ground level and one incision is made in the middle of the stock at depth of 2.5cm.

Select scion branch of the desired variety of 10 to 15 cm long having at least 3 to 4 buds. Prepare the scion by removing the leaves. Make a wedge shaped cut on the lower end of the scion of similar size as that of stock.

The cut surfaces of both the stock and scion are tied together with alkathene so that the cambium of each other comes in close contact.

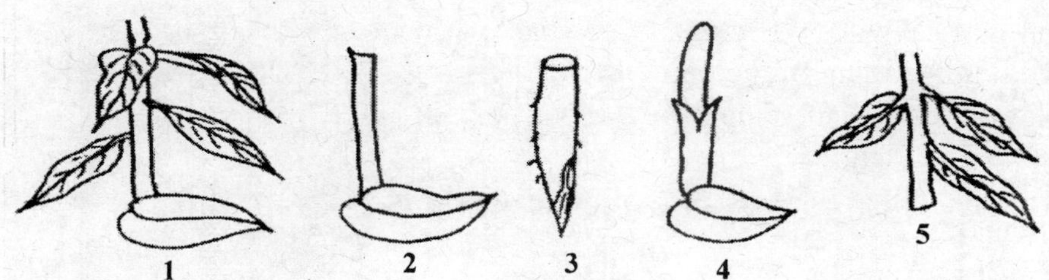

1. Root stock with cotyledon **2.** Root stock headed **3.** Scion in V cut with **4.** Scion wood

Soft Wood Grafting

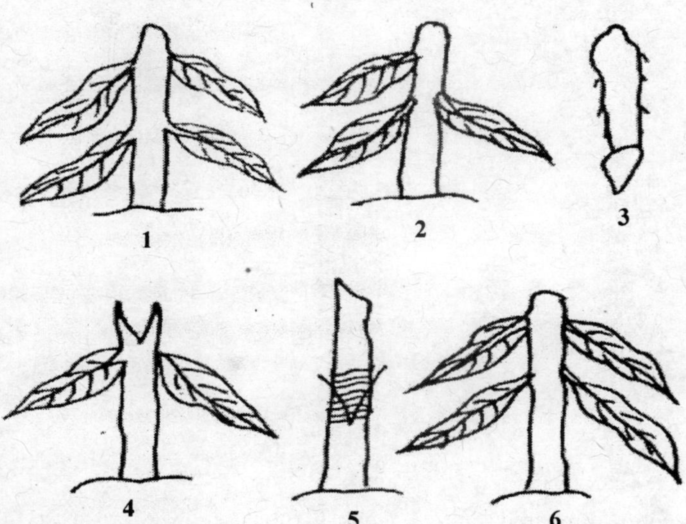

1. Root stock **2.** Root stock headed back **3.** 'V' shape cut on scion **4.** Scion inserted in stock. **5.** Scion wood. **6.** Prepared scion.

Plate 16:

PROPAGATION BY EPICOTYLE GRAFTING

Materials and equipment: Grafting knife, secateurs and tying material.

Procedure: Germinating seeds about 4 to 8 days old are used as root-stock when it is bronze in colour. Stock is removed at 5cm from the stone of ground level and one incision is made in the middle of the stock at depth of 2.5cm.

Select scion branch of the desired variety of 10 to 15 cm long having at least 3 to 4 buds. Prepare the scion by removing the leaves. Make a wedge shaped cut on the lower end of the scion of similar size as that of stock.

The cut surfaces of both the stock and scion are tied together with alkathene so that the cambium of each other comes in close contact.

Stone Grafting

1. Root stock with cotyledon 2. Root stock headed 3. Scion in V cut with 4. Scion wood

Soft Wood Grafting

1. Root stock 2. Root stock headed back 3. 'V' shape cut on scion 4. Scion inserted in stock 5. Scion wood 6. Prepared scion.

Plate 16.

15

Numerical on NPK in Orchard

Problem- Recommended dose of N:P:K for bearing mango tree 1.0 Kg N: 0.87 Kg P and 1.66 Kg K annually per plant. Available fertilizers are urea, DAP and MOP for supplying of NPK respectively.

So, for an orchard of 1 hectare area, the total number of plants is 100 if spacing is 10m x 10m

Therefore the total amount of N is 100 x 1.0= 100Kg.

P is 100 x 0.87 = 87 Kg.

K is 100 x 1.66=166Kg

Calculation

(A) 46 Kg P_2O_5 obtained from 100Kg DAP

1 Kg P_2O_5 obtained from 100/46

87 Kg P_2O_5 obtained from 100x87/46= 8700/46= 189.13Kg

(B) 100 Kg DAP contains 18Kg N

1 Kg DAP contain 18/100 N

189.13 Kg DAP contains 18x189.13/100= 3404.34/100 =34.04 Kg

Remaining amount of N (100-34.0=66Kg) supply by urea. Therefore for supplying 66 Kg N by 100 Kg urea because urea contains 46% nitrogen. Amount of Urea to supply = (100 x 66)/ 46 =143.5 kg Urea

(C) 60 Kg K_2O obtained from 100 Kg M.O.P

1 Kg K_2O obtained from 100/60Kg

166 Kg K_2O obtained from 100x166/60= 276.66Kg

Answer - Urea = 189.13 kg

Diamonium Phosphate = 34.04 Kg

Muriate of potash= 276.66 Kg

16

Mango Varieties

Important Commercial Mango Varieties

Northern Region: Langra, Dashehari, Chausa, Bombay Green, Amrapali, Sukul, Sipia

Eastern Region: Himsagar, Langra, (syn. Malda), Krishna-Bhog, Zardalu, Gulabkhas, Fazli.

Southern Region: Banganapalli, Bangalora, Mulgoa, Neelum, Swarnarekha,

Western Region: Alphonso, Kesar, Pairi.

Varieties suitable for rootstock:

(a) **For vigorous tree stature:** Bappakai, Olour, Chandrakaran

(b) **For salt tolerant:** Kurukkan, Bappakai, Neleshwar Dwarf

(c) **For dwarf tree stature**: Totapari Red Small, Nileswar Dwarf.

Langra

This is a chance-seedling. This is one of the most popular varieties and has a wide adaptability. Trees are tall, vigorous and spreading. This is a mid-season variety and biennial bearing habit. Excellent fruits quality having good acid/ sugar ratio but the keeping quality of fruit is poor. Fruits are medium to large in size and fibreless

Dashehari

This is also one of the excellent varieties because of its taste, flavour and pulp-fibre ratio. This variety also originated as a chance seedling in one of the orchard of Nawab of Lucknow. Trees are medium in size and not so vigorous and biennial bearing habit. It is a medium to big in size, good keeping quality, fruit quality, elliptical oblong in shape, greenish-yellow colour, rich in vitamin C and fiber-less.

Chausa

This is one of the sweetest and delicious mango variety originated as a chance seedling chausa village in Lucknow, Uttar Pradesh. Trees are medium, moderately vigorous and spreading. It has a biennial bearing habit. Fruits are medium to big in size, ovate to oblong in shape, yellow coloured, juicy flavour mild and poor keeping quality. It is a late maturing variety.

Bombay Green

This is one of the best early maturing mango variety. This had also originated as a chance seedling. The tree is medium, vigorous and spreading. Fruits are medium in size, ovate oblong in shape beak in absent, green-yellowish in colour, taste and flavour excellent, good keeping quality. This has biennial bearing habit and susceptible to malformation.

Himsagar

This is one of the best and popular varieties of West Bengal. This is a mid-season variety. Trees are medium, moderately vigourus spreading and biennial in bearing. Fruits are big in size, oval, beak absent, yellowish-green in colour, good fruit quality and good keeping quality.

Alphonso: (Syn- Haphus, Badam)

This is one of the excellent mango variety, which is exported to other countries to a large extent. Tree are medium to large, vigourus and upright. It has biennial bearing habit and mid season variety. Fruit are of excellent quality, medium size, ovate shape, attractive red coloured, very good keeping quality, good for canning excellent flavour and taste. In which they found TSS-17% and acidity-8.65% It has a very specific requirement and does well only on the West Coast of Maharashtra (Ratanagiri)

Krishnabhog

This is a late maturity variety. The trees are medium to large and moderately spreading. It is biennial in bearing and heavy yielder. Fruits are medium to large, round in shape, good in taste and have light yellow colour at the basal end of the fruit.

Gulabkhas

This is a variety from Bihar. Fruits have an attractive reddish colour. Trees are medium and moderately vigorous. It is a mid-season variety and biennial bearing habit. Fruits are oblong- oblique, small to medium, good taste, and keeping quality not so good.

Zardalu

This is a variety from Bihar. Due to its flavour of Zarda it occurs it name, trees are medium to moderate in size. It has a biennial bearing habit and fruiting in medium to moderate. Fruits are medium in size and mature towards the end of June. The colour of the fruit in apricot yellow, medium in size oblong.

Fazli

This is a late maturing variety. Fruits are very big, green in colour and fruit quality is poor. Trees are big, moderately vigorous and upright. It is biennial bearing in habit.

Banganpalli: (Syn.-Baneshan, Safeda)

It is an early maturity variety from south specially Andhra Pradesh. Trees are moderately vigorous and spreading. Bearing is moderate and fairly regular. Fruits are big in size, obliquely oval, golden-yellow in colour, smooth skin, good keeping quality, good fruit quality and pleasant flavour.

Bangalora: (Syn. Totapari, Collector)

This is also a widely cultivated variety of South. Bearing is heavy regular and an early maturing variety. Fruits are oblong, medium, beak prominent and out curved. The colour of the fruit in apricot-yellow, keeping quality good and fruit quality is poor.

Neelum

This is a regular bearing and heavy yielding variety of the South especially Tamil Nadu. Fruits are medium in size, ovate-oblique in shape and the beak in distinct. The colour of the fruit is orange-yellow, pulp soft, fibreless and quality is good. It is late maturing variety and keeping quality is good.

Kesar

It is one of the best varieties from Gujarat having biennial bearing habit. The trees are moderately vigorous and top rounded. Fruits are medium to large, oval, attractive light apricot yellow colour, taste very good, pulp little fibrous and sugar/ acid blend excellent keeping quality is not so good.

Pairi

This is also one of the best early maturing variety of Western and Southern India. Trees are large vigorous and spreading. Fruits are medium, ovate, attractive crimson shoulder on a yellowish-green background, fruit quality good, fiberless, taste sweet, good flavour and keeping quality not so good.

Some coloured varieties

India has a vast wealth of mango varieties. Some of the well known attractive coloured varieties

Alphonso, Arka Arun, Arka Puneet, Edward (Simmonds), Elcon, Gulabkhas, Haden, Irwin, Kesar, Pairi, Sensation, Sindhu, Husnara, La Mulagoa

Some Polyembryonic mango varieties

Kurukkan, Bappakai, Chandrakaran, Goa, Olour, Nilewar Dwarf, Bellary, Goa Kasargod, Salem. These are mostly confined to Southern state especially on the West Coast and are mostly wild. Almost all the cultivated mango varieties are monoembryonic.

SOME MANGO HYBRIDS

From I.A.R.I., New Delhi

1. **Amrapali** : It is a cross between Dashehari x Neelum, released from I.A.R.I., New Delhi in the year 1979.
2. **Mallika:** It is a cross between Neelum x Dashehari released from I.A.R.I., New Delhi in the year 1972.
3. **Pusa Surya:** Clonal selection from Cv. Elden.
4. **Pusa Arunima:** Amrapali x Red Sensation

Fruit Research Station, Sangareddy: (A.P.)

1. **AU-Rumani-** It is a cross between Rumani x Neelum.
2. **Manjira-** A cross between Rumani x Neelum

Marathwada Agricultural University, Parbhani:

1. **Niranjan-** Off season variety, salection at Himayatbag.
2. **Neelphanso-** Neelum x Alphonso
3. **Neeleshan-** Neelum x Baneshan
4. **Neeleshwari-** Neelum x Dashehari

Fruit Research Station, Periyakulam (T.N.)

1. **PKM-**1-A cross between Chinnasuvarnrekha x Neelum
2. **PKM-2** - A cross between Neelum x Mulgoa

Horticultural Experiment Station, Paria (Gujrat)

1. **Neelphanso-** It is a cross between Neelum x Alphonso
2. **Neeleshan -** A cross between Neelum x Baneshan
3. **Neeleshwar-** It is a cross between Neelum x Dashehari

Fruit Research Station, Vengurla, Maharashtra

1. **Ratna:** It is a cross between Neelum x Alphonso. It is free from spongy tissue.
2. **Sindhu:** Ratna x Alphonso

Fruit Research Station, Sabour (Bihar)

1. **Mahmood Bahar-** It is a cross between Bombai x Kalapadi
2. **Prabhashanker-** It is a cross between Bombai x Kalapadi.
3. **Rajendra Am -1** It is a cross between Langra x Sunder Prasad.
4. **Rajendra Am-1** (Alfazli)- It is a cross between Alphonso x Fazli

I.I.H.R., Bangalore

1. **Arka Arun**: It is a cross between Banganpali x alphonso
2. **Arka Anmol**: It is a cross between Alphonso x Janardan Pasand
3. **Arka Neelkiran:** It is a crcss between Neelam x Alphonso
4. **Arka Puneet:** It is a cross between Alphonso x Banganpali.

Fruit Research Station, Kodur, Andhra Pradesh

1. **Neeleshan** : Neelum x Baneshan
2. **Neelgoa:** Neelum x Perramulgoa
3. **Neeluddin:** Neelum x Himayuddin.

17

Propagation in Mango

Veneer grafting

It has one striking advantage in using the detached scion sticks for propagation, i.e. the shoot of the mother plant which is to be multiplied can be cut away from the mother plant and taken to the seedling growing in the nursery for grafting. Proper selection of scion is very important for the success of this method. The scion stick ought to be 3 to 4 months old and with some activated bud, either auxiliary or terminal. This is secured by defoliating the scion shoot about a week before the actual operation

This method can be practiced on seedling stocks grown in pots or nursery beds. A slanting cut about 5cm long on one side of scion stock, is given and the bark along with wood is removed, giving an oblique cut. Now a slanting cut on one side of the scion is made, which will just fit with the notch of the stock. The scion is then placed in position in such a way that the cambium rings of both stock and scion come in close contact. It is then wrapped tightly with 1.5cm wide tape of 200-300 gauge alkathene film keeping the terminal ends free. When the scion begins to grow at the top, (after about 3 weeks) the upper part of the stock is removed, thus forcing the buds to grow more rapidly. The plastic wrap is removed after 2-3 months.

The success is more or less the same in the entire scion (2.5 to 10.50 cm in length) but the growth of scion is always more when bigger scions are used.

Epicotyl grafting

Germinating seeds of about 4-8 days old are used as root stocks. The scions are prepared by prior defoliation of shoot of comparative thickness. Splice or wedge methods are used for grafting. For splice grafting the epicotyls is cut slantingly for 2-3 cm length and the lower position of scion is also cut to match it. The cut surfaces of both the stock and scion are tied together with alkathene film so that the cambium of each other comes into close contact. In wedge grafting, a vertical cut 4-6cm long is given into the beheaded epicotyl, to receive the wedge-shaped scion. This is then tied with alkathene film. The grafts prepared by these methods are planted immediately in pots and watered. Grafting is done in rainy season when there is high humidity in the atmosphere.

The scion sprouts within month of operation. The percentage of success in splice and wedge methods is 50.0 and 33.3 % respectively.

18

Insect, Pest and Diseases of Mango

MAJOR INSECT-PEST OF MANGO

1. Mango hopper (*Idioscopus atkinsani, I. niveosparsus, I. clypealis*)

This is one of the most serious pests of mango. They are small greenish wedge-shaped insect. They cause serious damage to the crop at the time of flowering and fruiting. Both adult and nymphs suck sap from the tender plant parts, inflorescence, young shoots and panicles. They excrete sugary shirring substance on which shooty moulds develops. Cloudy and humid weather are favourable for their development.

Control

i. **Neem products**: They act as repellent and antifeedant. Oil-based concentrate and kernel based concentrate (NSKE) @ 0.4% are effective

ii. Spray of Dimethoate@ 0.5 % or Carboryl, 0.2 % or Monocrotophos 0.04% or phosphamidom 0.05% will be most effective.

Mango Mealy Bug (*Drosicha mangiferae*)

This is another most serious pest. The insect can be easily recognised as the female bugs are white in colour, flat and elliptical. The nymph and adult female suck the sap from tender plant parts and inflorescence. Due to continuous sucking the plant parts dry up and causes considerable damage. The female lays in the soil around the tree trunk. After hatching then nymphs crawl up the tree.

Control

i. The soil around the tree trunk should be ploughed and raked during November and apply methyl parathion dust 2% 100g/tree or chloropyrphos 2% 100g/tree or Folidol (parathion methyl which will kill the nymphs.

ii. Application of neem cake is also effective

iii. Alkathere bands (400 gauge) of 20cm wide about 40-60cm above the ground is tied around the tree trunk. Little grease or mud is applied on the lower portion of the band.

iv. In case the above mentioned measures are not adopted and the bugs have ascended the tree spray of Monocrotophos (0,06 %) is effective,

3. Frult-fly (*Dacus dorsalis*)

This is another serious pest of mango fruit. The adult flies are dark brown in colour and lays eggs beneath the skin of ripening fruits. After hatching the maggots starts feeding on the pulp and forms a tunnel and makes the fruits unfit for consumption.

Control

i. The infested fruits should be collected and destroyed.

ii. Spray of carboryl (0.2 %) + molasses solution effective

4. Bark- eating Caterpillar (*Inderbela tetraonis*)

This pest is found in the old and neglected trees. The caterpillar enters deep inside the stem by feeding on the bark and other tissues disrupting the translocation the branch dry up and the tree become weak. The caterpillar can easily be detected by the presence of brown loose mass of the excreta at the mouth of the tunnel they make.

Control: The caterpillar can be killed by inserting a hard wire warped with cotton poured in Kerosene oil or ethylene glycol inside the tunnel followed by plagging its mouth with mud. Monocrotophcs (Nuvacron) or Nuvan D.05% may be placed inside the hole and it is plugged with mud paste.

Other insect pests are

1. Stem borer (*Batocera rufomaculata*)
2. Mango shoot-gall maker (*Apsylla cistellata*)
3. Shoot borer (*Chhumetia transversa*)
4. Scale insect (*Aspidiotus cestructor*)
5. Termites

MAJOR DISEASES OF MANGO

1. Powdery mildew *(Oidium mangiferae)*:- This is one of the serious disease affecting almost every variety. The disease appears when the temperature and humidity is more congenial. A white, powdery growth of the fungus appears on the leaves, flowers and fruits.

Control

i. Spray of wettable sulphur @0.2%

ii. Karathane @0.2%

2. Anthracnose (*Colletotrichum gloecsporioldes*): This disease is of wide spread occurrence, under favourable climatic conditions of high humidity cloudy weather and frequent rains, it mostly affects the tender parts of the tree such as young shoots, leaves, flowers and fruits. The diseases produces leaf spot, elongated necrotic patches, small dark spots on the leaves, panicle and fruit. Dieback, twig and fruit not are also common.

Control

i. Spray copperoxychioride 0.2%

ii. Spray Bordeaux mixture 3:3:60

iii. Spray Bavistin or Captan 0.2%.

3. Die Back (*Botryodiplodia theobromae*): The symptoms of this disease can be recognized by the complete drying of branches and twigs generally from the upper portion leading to its death. The colour of the bark darkens and discoloured. The affected portion advances and the whole branch or the twig starts withering. The disease spread from top of the twig or branch downwards

Control

i. Use of the healthy scion and sterilized grafting knife in used

ii. Pruning of the diseased portion up-to 10-15cm below and spraying of copper oxychlorice 0.2% or Bordeaux mixture 5:5:50.

4. Bacterial Canker (*Xanthomonas compestris*): The disease attack leaves twigs, branches and fruits. The disease first appears as water of soaked irregular dark brown spots but latter on the spots gradually increase in size and become slightly raised. Several lesions coalesce to form necrotic cankerous patches. In severe infection the leaves and fruits drop off.

Control

i. Proper orchard sanitation should be adopted

ii. Spraying of streptomycin 0.2% or Agromycin 0.2% or streptocycline 0.2 % or Bavistin 0.2% has been found effective

Other diseases are

1. Leaf blight (*Macrophoma mangiferae*)
2. Phoma blight (*Phoma glomerata*)
3. Sooty mould (*Capnodium ramosum*)

19

Training and Pruning of Grape Vine

The following special terms are used in grape cultivation:

1. **Training:** Consists chiefly in giving a shape to the vine to the various forms of support.
2. **Pruning:** Consists in removing canes, shoots, leaves diseased, unproductive and other vegetative parts of the vine.
3. **Thinning:** Thinning is the removal of flower cluster, immature fruit cluster or parts of cluster.
4. **Shoots:** Shoots are the current season's top vegetative (succulent) growth arising from a bud after pruning.
5. **Canes:** Canes are the woody matured shoots.
6. **Trunk:** Trunk is the main stem of the plant or the undivided body of the vine.
7. **Arms:** Arms are the primary; secondary or the tertiary branches.
8. **Spur:** Spur is the basal portion of the cane having one to four buds or nodes.
9. **Renewal spur and foundation spur:** The spur having one bud which remains after pruning. It is called the renewal spur or the foundation spur as it forms the foundation on which the next years cane and both growths are borne.
10. **Fruit cane:** It is the basal portion of a cane, 8-16 buds long, are used to produce the crop on care pruned vines, t is always removed at the time of pruning.
11. **Water sprouts:** It is any shoots that arise on any part of a vine older than one year.
12. **Suckers:** It is water sprouts that arise below ground from the root zone.

13. **Girdling:** Removal of narrow ring of barks about 3-5 mm from the trunk and arms. Girdling improves the berry size and advances berry maturation. Girdling of trunk affects the entire vine.

14. **Pinching:** It is done by removing the succulent shoot tip (2-3cm).

20

Training in Grape Vines

The various systems of training the vines adopted in the vine-growing centres in India are as follows:

1. Head system
2. Bower system
3. Kniffin system
4. Overhead trellis

1. **Head System:** It is the easiest and cheapest system of training vines like a dwarf bush. The vine is allowed to grow on a single stem with the help of a stake (Fig. 1). The vines are supported by a vertical stake. These stakes are fixed close to the vine. The vines after attaining a height of 1.0m to 1.2m, it is cut back to produce side shoots. After keeping 2-4 lateral branches in all directions, 75cm above the ground, the rest of the shoots are thinned out. The lateral branches are pruned to 2 buds at the time of pruning, which will produce secondary arms. Generally, 2-3 arms of about 20-30cm are kept on each lateral. Canes are developed from these laterals. About 8-10 canes can be developed on each secondary lateral. After about 3-4 years the vine becomes a dwarf bush. The vines are planted at a distance of 1.8mx2.0 m apart or even closer at a distance of 1.8mx1.5m apart.

   The advantage of this system is that it s less expensive and simple. Intercultural operations can also be done easily. But production is less as compared to other systems.

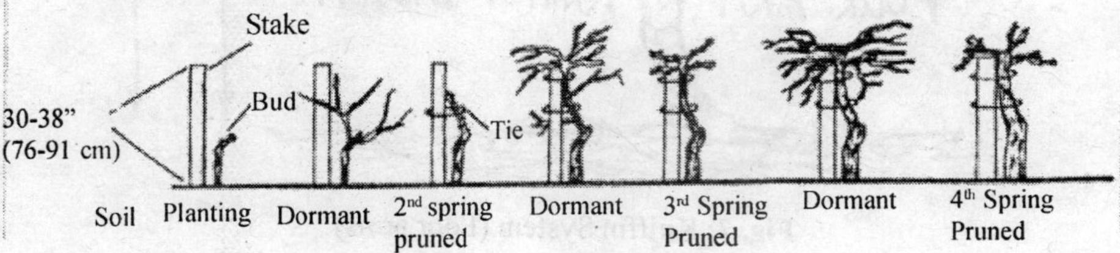

Fig. 1: Head System

2. **Bower system:** This system is also called arbour or pergola. This system has been formed to be the most suitable for almost all the varieties grown commercially in India. This system is best suited for vigorous varieties. In this system the vines are spread over wire netting at a height of 2 to 2.5 m above ground. The whole wire netting is supported by strong pillars made up of concrete, brick or poles. Vine is allowed to grow in the form of single shoot till it reaches the wire netting. All lateral branches are removed. The vine is supported by bamboo sticks ties with a string. When the vine reaches the netting, it is pinched off so that side shoots are produced close to the netting. Later, two strong branches are allowed to grow on opposite directions the wire level for training as primary arms, all others are removed. These are allowed to grow up-to the space provided. On each primary arms three secondary arms at a distance of 50-60cm apart are allowed to develop. On secondary arms, shoots are allowed to develop at 15-20cm apart. The tertiary shoots are proned to 2 to 3 buds in the following season. The new shoots arising from these spur will provide the fruiting and renewal spurs for the following year. This system provides large area on the vine for fruiting, which results in higher yield.

   According to variety vines are planted at a distance of 3 m x3m or 4.5m x 3.0m. The only disadvantage of this system is its high cost for installation.

3. **Kniffin system:** This system of training grape vines was developed by Mr. William Kniffin of New York in 1850. In this, system two wire trellis are strong horizontally from two vertical posts (Fig. 2). The lower wire is placed one metre above the ground and the upper wire half metre above the lower wire The single stemmed vine is allowed to grow forming

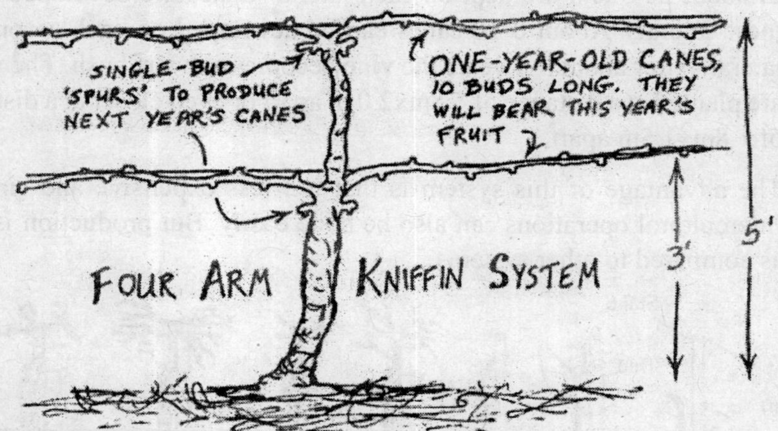

Fig. 2: Kniffin System (Four arms)

four arms, two arms on either sides of the trunk, two at lower wire and two arms at upper wire on both the sides. The pruning of vine consists in cutting back the canes to 6-8 buds. The bearing shoots are allowed to hang on the wire. The main trunk is tied to each wire and the canes are tied along the wires. There are several modification made in the original Kniffin system. There may be three horizontal wire trellies at different heights. This system is suitable for varieties like, Bhokari, Thompson Seedless and Kandhari, whose buds are comparatively less fruitful near the base than located further, up on the cane.

4. **Overhead trellis:** This system is also known as telephone system (Fig. 3). This system was first introduced in Poona by Dr. N. Gopal Krishanan in 1960. In this system the vine is allowed to grow straight up-to a height of 1.5m and then trained overhead on a canopy of 3-5 wires fixed to the cross arms supported by vertical pillars or posts at 40-50cm apart. The vines after reaching off to encourage side shoots near the wires. Two vigorous shoots are selected as primary arms from which four vigorous laterals on each side along the wire are allowed to develop as secondary arms. The canopy of wires carries the fruit well above the ground levels. The bunches remain away from the sun resulting in production of superior quality berry. This system is suitable for hotter regions. The vines are planted at a distance of 3mx3m.

Fig. 3: Overhead trellis

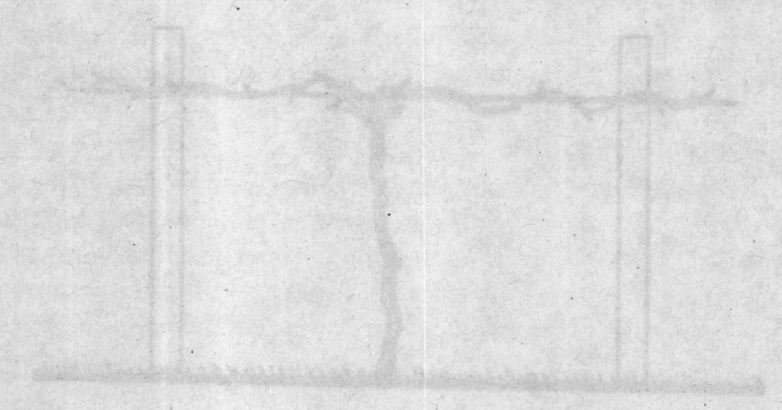

21

Plant Growth Regulators in Fruit Crops

Growth

Growth means quantitative increase in the planwt body e.g., increases in the length of stem, roots, number of leaves, fresh and dry weight of fruits etc.

Development

It refers to the qualitative changes from the initiation of growth to the death of a plant or plant parts like seed germination, formation of flowers, fruits and seeds, emergence of buds, leaves and falling of leaves and fruits.

Plant growth regulators

PGRs are organic compounds other then nutrients, which require in small amounts for promotes, inhibits or modify the physiological processes in plants.

Hormones

It may be defined as substance synthesized in one part of the plant and translocated to another part, where they exert their effect i e., the site of production and site of action are different. It may be defined growth hormone as substances which are synthesized in particular cells and transferred to other cells where in extremely small quantity influence development process.

Hormones involved in growth are called growth hormones.

There are different categories of plant growth substances, which may be broadly classified into growth promoting and growth retarding substances or into naturally occurring growth substances and synthetic growth substances. However, now a days the term growth regulating substance is preferred because it include both naturally and synthetic growth substances. Growth regulating substances may be classified into following categories.

1. Auxin

It is growth regulators, which at low concentration (<.001m) promotes growth along longitudinal axis of shoots, freed as far as possible from their own inherent growth promoting substance.

i. Indole auxins - IAA, IBA, IPA.

ii. Nepthyle groups - NAA, NOA

iii. Phenoxy group - 2,4-D; 2-4,5-T

iv. Benzoic group - 2,4,6- Trichlorabenzoic acid, 23,6-trichlobenzoic acid

2. Gibberllins

GA are hormones, which promotes plant growth specially stem.

Gibberellic acid – GA_1 - $GA_{120.}$ Young leaves are the site of active GA synthesis. GA synthesis also occurs in root, embryo and endosperm.

i. It promotes seed germination

ii. Cell elongation in intact plants

iii. Help in breaking rest and dormancy

iv. Prevent flower initiation.

v. Induce partherocarpic fruit set

3. Cytokinin

The cytokinins are substances, which primarily act on cell division and have little or no effect on extension growth e.g., Benzyle adenine, Kinetin, Zeation etc.

4. Ethylene

Gases present in smoke could modify plant growth. Neljubor (1901) was the first to show the importance of ethylene present in illuminating gas as a growth regulator of plants. Virtually all living plant cells are capable of producing it. An unique feature of this plant hormone is that, it is regularly flushed out of the plant.

i. It promote germination of seeds.

ii. Seed dormancy is overcome

iii. At higher concentration growth of the sprout is checked.

iv. It hastens the ripening of fruits

v. Accelerate the latex flow from rubber plant.

5. Inhibitor

i. It may be natural as well as synthetic e.g., ABA, MH, Dormin, SADH.

ii. It suspresses the growth of apical meristem.

iii. Higher concentration of inhibitors may cause malformation

6. Retardents

i. Retardents only slow down the growth of the apical meristem.

ii. Higher concentration of retardants does not cause any maformation to the plant e.g. CCC, Phosphon-D, AMO-1618, Alar, Phos-S.

7. Flowering Hormones

(i) Florigen

(ii) Anthesin

Section B
Vegetable Production

22

Kitchen Garden

Kitchen garden

Growing of vegetable crops in the backyard of residential houses to meet the requirements of the family all the year round. It is most ancient type of vegetable farming. Such type of gardening is practiced in cities where land is the limiting factor. Vegetables can also be grown on the roof of buildings if land is not available.

Aim of kitchen garden

The main aim is a kitchen garden in the maximum output and a continuous supply of vegetables for the table throughout the year.

Principles

Following certain principles in the layout of kitchen garden, the above aim can easily be fulfilled:

1. The perennial plant should be located on one side of the garden, usually on the rear end of the garden so that they may not shade other crops, complete for nutrition.
2. Adjacent to the foot path all around the garden and the central foot path may be utilized for growing different short duration green vegetables.
3. The fence or trellises around the home garden may be utilized for growing light creepers like Basella, Coccinea, Sponge-gourd and Bitter-gourd.
4. The compost pit are placed in two corners of the garden. They are meant for garden waste and kitchen wastes.
5. Pandals may also be erected over the central foot path, grapes varieties Anab-e-Shahi or Black Prince may be trained over it.
6. Both the sides of the central foot path may be utilized to train plants.
7. The bunds separating the beds may be used for growing root crops or onion.

8. The conveniently divided small plots may be utilized to produce as much as possible by following a very intensive method of cultivation.

Advantages of kitchen gardening

1. Kitchen gardening a best means of recreation and exercise in spare time for all the family members.
2. There is saving of money as fresh vegetable can be available with less expenditure.
3. It may ensure enough vegetable for all the people at less production cost.
4. Fresh and nutritious vegetables are available whenever required.
5. The kitchen gardeners feel most satisfaction by consuming their self grown vegetables.
6. More nutrients can be assured in daily diet by increasing the availability of nutritious vegetable.

Constraint of kitchen gardening

1. Land is the limiting factor in cities and town for sparing kitchen gardening.
2. Shade and lack of irrigation facilities some time limit the adoption of kitchen gardening.
3. Lack of awareness among the kitchen gardening about vegetable cultivation.

Model kitchen garden and the cropping arrangement

Our dietician recommends under our condition to consume 300gms of vegetables daily by an adult and based on this, a kitchen garden should supply 1.5Kg of fresh vegetables to an average family size of foure adults and two children. This quantity of fresh vegetable can be assured from a kitchen garden, laid out in an extend of 250 square meters following the above principles

It is advisable to make a plan before undertaking the planting of a garden. The location of plots, crops to be grown, the probable date of planting, spacing between the plants, intercropping and succession planting should be clearly indicated in the plan.

A cropping pattern which may be helpful for kitchen garden under Uttar Pradesh condition is suggested below:

It may be observed from the above crop arrangements that throughout the year some crop is grown in each plot without a gap between any two crops (succession cropping) and wherever possible two crops (one long duration and the other short duration one) are grown together in the same plot (companion cropping):

Plot No.	Name of vegetable	Season
1.	Tomato and onion	June-September
	Radish	October-November
	French bean	December-February
	Bhindi	March-May
2.	Cauliflower	September- February
	Cowpea (Summer)	March-May
	Cowpea (Rainy)	June-August
3.	Brinjal with spinach as a intercrop	June- September
	French bean	October- December
	Tomato	June- September
4.	Bhindi and Radish	June-August
	Cabbage	September- December
	Cluster bean	January- March
5.	Cauliflower (mid season)	July-November
	Radish	November- December
	Onion	December-June
6.	Potato	November- March
	Cow pea	March-June
	Cauliflower(early)	July-October
7.	Lab-lab (Bush type)	June-August
	Onion and Chilli	September- December
	Bhindi	January- March
	Coriander and Amaranthus	April-May
8.	Cucumber	July- September
	Potato	October- January
	Cowpea (Half plot)	February- May
	French Bean (Half plot) or	February- June
	Summer squash (Half plot)	February- June
9.	Brinjal	July – December
	Broccoli	December- March
	Cucumber	April- June
10.	Bitter gourd (Half plot)	July- November
	Bottle gourd (Half plot)	July- November
	Fenugreek	November- March

Perennial plot (quick growing fruit tree/vegetables for kitchen gardening):

Drumstick : One row

Banana (Culinary cultivation) : Three rows

Papaya	:	Two plant each of different variety
Kagzi lime	:	One plant
Guava	:	One plant
Mango (Amrapali)	:	One plant
Tapioca	:	Two rows
Curry leaf	:	One row
Asparagus	:	Two rows

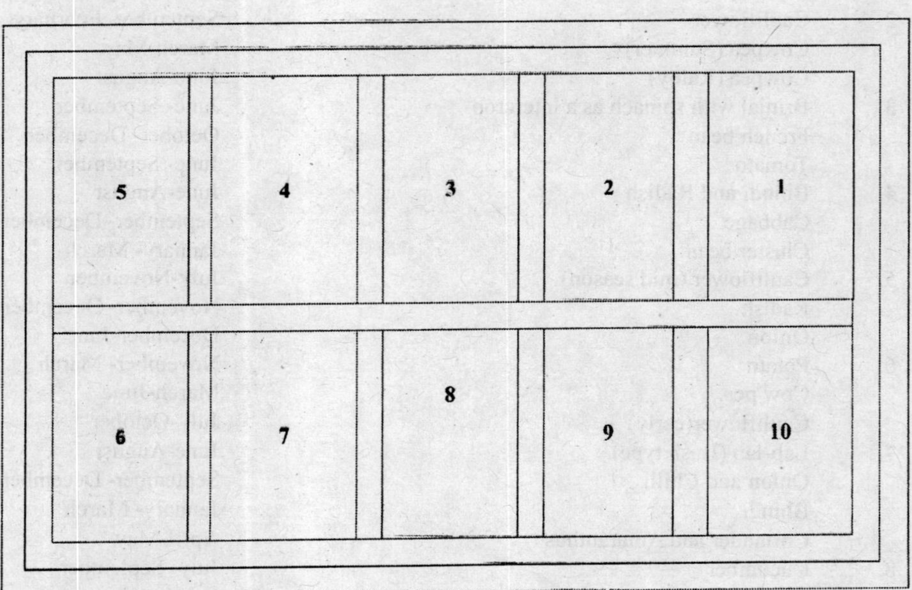

Layout of kitchen garden

23

Numerical on Kitchen Garden

Problem 1: Prepare a lay-out plan of model kitchen garden for a family having five (5) members. Calculate the total amount of vegetable required per year and area of the same kitchen garden. Give the suitable crop rotation and show by diagrammatic representation.

Ans.: Total amount of vegetables required annually

Generally 300 grams/capita/day vegetables are required.

So that, 5 members required quantity = 5x300 1.5Kg

So that, annually required vegetables = 365x1.5 = 547.5 Kg

Area of kitchen garden

Per hectare yield of vegetable= 225q = 22500 kg,

22500Kg vegetable produced in the area of = 10000 m²

So that 1kg " " " " " = 10000/22500

547.5 kg " " " " " = (10000/22500)x 547.5

= 243.33

= 250 m²

Layout of kitchen garden

Area 250m² (length = 25m and width = 10m)

(A) Area under Length)

(i) 2 sides barbed wire (1/2 m each) = 1 m

(ii) 4 Plots (5m each) = 20 m

(iii) One plot = 4 m

Total area = 25 m

(B) Area under width

(i) 2 sides barbed wire (1/2m each)	=1m
(ii) 2 irrigation channel (.25m each)	=0.5m
(iii) Main road (1/2)	= 0.5m
(iv) Four plots (each 2m)	=8.0m
Total area	=10.0m

Crop Rotation

Plot No.1 - Chilli (15 June- 15Feb)
Radish (20 Feb-5 April)
Leafy vegetables (10April-5May)

Plot No.2 - Cowpea (25 June- 25 Feb)
Potato (15 Oct-20 Jan)

Plot No.3 -Cauliflower+ Radish (30June-10Sep)
Cauliflower- (15 Sep- 25 Dec.)
Okra (10 Jan 15Cct)

Plot No.4 -Turnip (1 July-15Oct)
Onion (20 Nov.-15 March)
Bitter gourd (20 March -25 June)

Plot No.5 -Okra (25 June-30 Sept.)
Sugarbeet+ Turnip (15 Oct-30Jan)
Tinda (15 Feb.-15June)

Plot No.6 -Clusterbean (1 July- 30 Oct.)
Carrot+ Radish (15 Nov.- 20 March)
Palak (30 March-25June)

Plot No.7 -Brinjal and Its rattcon (25 June- 10 June)

Plot No.8 - Tomato (30 March-25 June)
Cowpea (1 July-30Jan).

Plot No.9 - Leafy vegetables (15 Feb-30 May)

Spongegourd (10 June-20Oct)

Garlic (30 Oct- 10 Feb)

Plot No.10 - Sagarbeet (15 Aug.-15 Oct)

Long melon (20 Feb.-15 April)

Colocasia (20 Apr. -5 Aug.)

Plot No.11- Tinda (15 July- 150ct)

Pea (20 Oct-20March)

Watermelon (25 March-30June)

Plot No.12 - Okra (15 Feb. -15July)

Caulilower (1 Aug.- 10November)

Onion (15 Nov.-10Feb)

Plot No.13 - Sponge gourd (15July- 15 Nov.)

Coriander green (20Nov. -25Jan)

Bittergourd (5 Feb.-30June)

Plot No.14 - Maize (1 July -10 Nov.)

Potato (15 Nov-10 March)

Muskmelon (15 March-30June)

Plot No.15 - Chilli (25 June- 15 Nov.)

Potato (20Nov.-15 March)

Tinda (20 March- 20June)

Plot No. 16- Kagzi lemon

Plot No. 17- Papaya

Plot No 18 -Banana

Plot No. 19 - Karonda

Plot No. 20- 1/2 Part Phalsa +1/2 Part compostpit

Problem 2- Prepare a layout plan of a kitchen garden for a family of 6 (4 adult and 2 child) members calculates the total amount of vegetable requirement per year and area of the kitchen garden. Illustrate it by a diagrammatic presentation.

Ans.

Amount of vegetable required /day by a single person- 300g

Vegetable requirement for (6 members) 4 adult = 300x4= 1200g

2 child = 150x2 = 300g

Total vegetable requirement for all members = 1200+300= 1500g or 1.5kg

Total Vegetble requirement by a family of 6 in one year = 1.5 x365=547.500

Average production of vegetable/ha=25,000Kg.

1 hectare=10,000 Sq. m

For producing 1 Kg vegetable in the area of = 10,000/25,000

For producing 547.5 Kg vegetable in the area of =10, 000x547.500/25,000.

=219 m^2 or 220 m^2

Area of the kitchen garden length x breadth 20m x 11m

a) i. Area in Length= 20m

ii. 4 plots of 4m= 16m

iii. 1 plot of 3m = 3m

iv. Fencing on two sides (0.5m) = 1.0m/20m

b) i. Area in breath = 11m

ii. Four plots of 2m = 8m

iii. Two irrigation channels (0.5m) = 1.0m

iv. Two sides fencing (0.5m) = 1.0m

v. Two sides paths (0.5m) = 1.0m/11m

Fencing
Path
FENCING
PATH
5 4 3 2 1
6 7 8 9 10
Irrigation channel
Main rode
Irrigation channel
11 12 13 14 15
16 17 18 19 20
PATH
FENCING
Path
Fencing

Layout of Model Kitchen Garden

Designing the kitchen garden

While designing a kitchen garden various factors such as availability to land situation, labour, soil, availability of water etc. should be taken care off

Location and size of kitchen garden

1. Kitchen garden is generally located at the backside of the house.
2. All the waste water from the kitchen can be utilized in the kitchen garden.
3. Vegetable crops should be located away from the shade of the trees.

Size of kitchen garden depends upon

1. Availability of land.
2. Daily requirement of vegetables in the family.

Economic utilization of space

1. Successive sowing of vegetables should be done for continuous supply of vegetable.
2. Various types of vegetables should be sown in different parts of the garden for different seasons

3. The plots should be divided with proper ridges and paths.
4. The ridges which separate the beds should be utilized for growing root crops like radish, carrot turnip, beet etc
5. The layout of the garden should be such that it looks attractive and allow access to all the parts easily.
6. The inter-space of some crops, which are slow growing like brinjal, cauliflower, cabbage quick growing crops like turnip, radish, spinach, lettuce etc.
7. One compost pit can be dug in one corner of the kitchen garden.
8. Some perinnial quick growing fruit crops like papaya, kagzi lime, banana, guava etc. can be planted on one side of the garden preferably on northern side, so that they may not shade other crops.
9. Climbing type vegetables like cucurbits, sem etc. may be trained on the fence.

24

Seed Extraction of Tomato

In tomato seed extraction is the very important operation. Tomato seed can be extracted by following way.

(1) **Fermentation methods**: Red ripe fruits should be picked and crushed in non metallic pots. Profuse foam formation indicates that fermentation is completed. Fermentation of seed material should not be prolonged beyond 72hr to get seed of high germination. Preferably the seed should be washed free of pulp after 48hr require temperature uses.

(2) **Acid treatment methods**: In case of urgency seed may be washed by using 1.0 to 1.5 liters of 35 to 38% concentrated HCL/g tomato fruit pulp. Fermentation is completed in about 30 minutes. Clean seed are obtained by stirring and washing procedure are quickly as possible.

(3) **Alkali treatment methods**: In case Alkalie 10% sodium bicarbonate is mixed in equal volume of pulp or 300 g washing soda dissolve in 4 liter of boiling water is sufficient for 4 liter fruit pulp.

(4) **Mechanical methods**: It is used for large scale operations put the ripe fruits into the mechanical seed extractor for crushing and separation of the seed and gel from the pulp. By acid 0.8% solution of acetic acid for 24hr at or below 21^{0}C (1kg seed/10 liter of acetic acid).

Seed drying:- Seed are dried in the sun in case of temperature exceeds 40^{0}C the seed drying under partial shad is desirable. Seed may be dried in seed dryer's moisture level 6-8%.

Seed yield- The average seed yield 80-100 kg/ha

Seed treatment to remove viruses: By 30 minutes in 10% solution house hold bleach (0.525% sodium hypochlorite Nacl) and 5hr in 5% Hcl.

Seed standard and seed tagging: In which Foundation Seed and Certified Seed are minimum germination 70% and minimum purity 98% without any admixture of other crop seed.

25

Cost Benefit Ratio of Vegetable Crops

Problem: Cost of cultivation depending upon the seed and time. More yields can be obtained during right time of sowing, but net profit is very less due to the glut in the market. We can get more return from early or late sown crop during both the season due to the high cost in the market.

Ans.

A. Cost benefit ratio of Tomato

S.No.	Operations / Items	Quantity	Rates	Input (Rs.)
A. Common cost				
1.	Field preparation			
A	Ploughing by disc harrow	2	400/ha/ Ploughing	800=00
B	Ploughing by cultivator	2	300/ha/ Ploughing	600=00
C	Planking	2	50/ha/Planking	100=00
2.	Seed Cost (a) seed	250g (hybrid)	20000/kg	5000=00
	(b)Sowing cost	2lab	80Rs./lab/day	160=00
3.	Lay out and transplanting	40lab	80Rs./lab/day	3200=00
4.	Application of fertilizer			
	(a) Nitrogen 120kg through Urea	260.40kg	5Rs./kg	1302=00
	(b)Phosphorus 60kg through SSP	375kg	4Rs./kg	1500=00
	(c) Potash 50kg MOP	83.50kg	5Rs/kg	418=00
5.	Irrigation (5times) through tubewell	25 Hours	50Rs./hour	1250=00
	(a) Irrigation hours			
	(b) Labour for irrigation	5lab	80Rs/lab/day	400=00
6.	Intercultural operation			
	(a)Two weeding	30lab	80Rs./lab/day	2400=00
	(b)Two Hoeing	30lab	80Rs./lab/day	2400=00
	(c)Two earthing	30lab	80Rs./lab/day	2400=00
7.	Plant Protection			
	(a) 2 spray of Deithane M-45	2 lit	192Rs./lit	384=00
	(b) 2 spray of Mnocrotophos	1lit	350Rs./lit	350=00
	(c) Spray Cost	4lab	80Rs./lab/day	320=00
8.	Harvesting	40lab	80Rs./lab/day	3200=00
9.	Rental value of land	4month	2400 Rs./ha/year	800=00
10.	Miscellaneous charge @ 5% of 26984			1350=00
			Total	26984=00
			Grand total	28334=00

(A) Production = 350q = 350x100 = 35000kg

(B) Selling price = 3Rs/kg

(C) Net income = 35000 x 3 = 105000/=

(D) Net profit = 105000-28334 = 76666=00

(E) Cast benefit ratio = 1:2:70

B. Cost benefit ratio of Cucurbitaceous crops:

S.No.	Operations / Items	Quantity	Rates	Input (Rs.)
A Common cost				
1.	Field preparation			
A	Ploughing by disc harrow	2	400/ha/ Ploughing	800=00
B	Ploughing by cultivator	2	300/ha/ Ploughing	600=00
C	Planking	2	50/ha/Planking	100=00
2.	Layout preparation	20labs	80Rs./lab/day	1600=00
3.	Cast of seed	3.5kg	250Rs./kg	875=00
4.	Application of fertilizer			
	(a) Nitrogen 80kg through Urea	130.20kg	5Rs./kg	651=00
	(b) Phosphorus 50kg through SSP	250kg	4Rs./kg	1000=00
	(c) Potash 40kg MOP	66.40kg	5Rs/kg	332=00
	(d) FYM	10tonnes	200/tonnes	2000=00
	(e) Application charge	6lab	80Rs./lab/day	480=00
5.	Irrigation (5times) through tubewell (a) Irrigation hours	20 Hours	50Rs./hour	1000=00
	(b) Labour for irrigation	5lab	80Rs/lab/day	400=00
6.	Intercultural operation			
	(a) 1 hand weeding	25lab	80Rs./lab/day	2000=00
	(b) Two Hoeing	15lab	80Rs./lab/day	1200=00
	(c) 1 earthing	30lab	80Rs./lab/day	1200=00
7.	Plant Protection			
	(a) 2 spray of Deithane M-45	2 lit	192Rs./lit	384=00
	(b) Application of furadon	3kg	63Rs/kg	189=00
	(c) Insecticides	500ml	670Rs/lit	385=00
	(d) Sevin	4kg	50Rs/kg	200=00
	(e) Spray Cost	4lab	80Rs./lab/day	320=00
8.	Labour charge for topdressing of urea	4lab	80Rs./lab/day	320=00
9.	Harvesting	12lab	80Rs./lab/day	960=00
10.	Rental value of land	4month	2400 Rs./ha/year	800=00
			Total cast	17796
11.	Miscellaneous charge @ 5% of 17796			890=00
			Grand Total	18686=00

(A) Production = 250q = 250x100 = 25000kg

(B) Selling price = 2Rs/kg

(C) Net income = 25000x2 = 50000/=

(D) Net profit = 50000-18686 = 31314=00

(E) Cast benefit ratio = 1:1:67

C. **Cost benefit ratio of Cauliflower and Cabbage:**

S.No.	Operations / Items	Quantity	Rates	Input (Rs.)
A	**Common cost**			
1.	Field preparation			
A	Ploughing by disc harrow	1	300/ha/ Ploughing	300=00
B	Ploughing by cultivator	3	400/ha/ Ploughing	1200=00
C	Planking	2	50/ha/Planking	100=00
2.	Layout preparation and transplanting	40labs	80Rs./lab/day	3200=00
3.	Cast of seed (a)seed	450g	15000Rs./kg	6750=00
	(b)Sowing charge	2lab	80Rs./lab/day	160=00
4.	Application of fertilizer			
	(a) Nitrogen 150kg through Urea	326kg	5Rs./kg	1630=00
	(b) Phosphorus 100kg through SSP	625kg	4Rs./kg	2500=00
	(c) Potash 100kg MOP	167kg	5Rs/kg	835=00
	(d) FYM	20tonnes	200/tonnes	4000=00
	(e) Application charge	6lab	80Rs./lab/day	480=00
5.	Irrigation (5times) through tubewell	25 Hours	50Rs./hour	1250=00
	(a) Irrigation hours			
	(b) Labour for irrigation	5lab	80Rs/lab/day	400=00
6.	Intercultural operation			
	(a) 2 weeding	40lab	80Rs./lab/day	3200=00
	(b) Two Hoeing	40lab	80Rs./lab/day	3200=00
	(c) 1 earthing	20lab	80Rs./lab/day	1600=00
7.	Plant Protection			
	(a) 2 spray of Deithane M-45	2 lit	192Rs./lit	384=00
	(b) (b) 2 spray of Mnocrotophos	1lit	350Rs./lit	350=00
	(c) Spray Cost	4lab	80Rs./lab/day	320=00
8.	Harvesting	40lab	80Rs./lab/day	3200=00
9.	Rental value of land	4month	2400 Rs./ha/year	800=00
	Total			35859=00
10.	Miscellaneous charge @ 5% of 35859			1793=00
	Grand Total			37652=00

(A) Production = 350q = 350x100 = 35000kg

(B) Selling price = 3Rs/kg

(C) Net income = 35000x3 = 105000/=

(D) Net profit = 105000-37652 = 67384=00

(E) Cast benefit ratio = 1:1:78

D. Cost benefit ratio of Okra:

S.No.	Operations / Items	Quantity	Rates	Input (Rs.)
A	**Common cost**			
1.	Field preparation			
A	Ploughing by disc harrow	1	400/ha/ Ploughing	400=00
B	Ploughing by cultivator	2	300/ha/ Ploughing	600=00
C	Planking	2	50/ha/Planking	50=00
2.	Layout preparation	35labs	80Rs./lab/day	2800=00
3.	Cast of seed (a)seed	10kg	250Rs./kg	2500=00
	(b)Sowing charge	8lab	80Rs./lab/day	640=00
4.	Application of fertilizer			
	(a) Nitrogen 60kg through Urea	130.20kg	5Rs./kg	651=00
	(b) Phosphorus 50kg through SSP	312.50kg	4Rs./kg	1250=00
	(c) Potash 40kg MOP	66.40kg	5Rs/kg	332=00
	(d) FYM	10tonnes	200/tonnes	2000=00
	(e) Application charge	4lab	80Rs./lab/day	320=00
5.	Irrigation (5times) through tubewell	20 Hours	50Rs./hour	1000=00
	(a) Irrigation hours			
	(b) Labour for irrigation	5lab	80Rs/lab/day	400=00
6.	Intercultural operation			
	(a) 1hand weeding	30lab	80Rs./lab/day	2400=00
	(b) Two Hoeing	20lab	80Rs./lab/day	1600=00
	(c) 1 earthing	20lab	80Rs./lab/day	1600=00
7.	Plant Protection			
	(a) 2 spray of Deithane M-45	2 lit	192Rs./lit	384=00
	(b) Application of furadon	4kg	75Rs./kg	300=00
	(c) Spray Cost	5lab	80Rs./lab/day	400=00
	(d) Top dressing cost	4lab	80Rs./lab/day	320=00
8.	Harvesting	15lab	80Rs./lab/day	1200=00
9.	Rental value of land	4month	2400 Rs./ha/year	800=00
	Total			26984=00
10.	Miscellaneous charge @ 5% of 26984			1350=00
			Grand Total	28334=00

(A) Production = 250q = 250x100 = 25000kg

(B) Selling price = 3Rs/kg

(C) Net income = 25000x3 = 75000/=

(D) Net profit = 75000-23097 = 48097=00

(E) Cast benefit ratio = 1:2:08

26

Nursery Management of Vegetable

Nursery

An area where the planting materials are raised for sowing or planting in garden or fields, in other words, the nursery industry involves the production and distribution of different kind of planting material.

Some of the vegetables sown directly in the field are Okra, Pea, Cowpea, Frenchbean Dolichosbean, Radish, Carrot, Bottle gourd, Bitter gourd, Ridge gourd, Spongegourd, Ashgourd, Muskmelon, Watermelon, Cucumber, Pumpkin, Roundmelon, Spinachbeet, Spinach, Coriander, Amaranthus, Lettuce, Fenugreek etc. Some vegetables are propagated through vegetative means like Colocasia, Ginger, Garlic, Potato, Sweet potato, Drumstick, Asparagus, Elephant foot, Pointedgourd and Ivygourd. But there are certain vegetables having very small seeds and first sown in the nursery for better care and to combat the main field. Such vegetables are Tomato, Brinjal, Chilli, Capsicum, Broccoli, Endive, Chicory, Cabbage, cauliflower, Khol-Khol, Kale, Parsley, Lettuce etc.

Raising of seedlings in the nursery beds is economic as well as easier to take care of young, tender seedlings against disease, insect pests and weeds. Better combat against biotic stress like bright sun, rains and low temperature and economisation by saving land, labour and capital to fetch higher market price.

Selection of site and preparation of soils for seed bed

While selecting a site nursery beds, fallowing points should be considered. The area should be free from water logging, away from shade, desired sunlight, near to water supply and should be fence from pet and wild animals. Sandy loam to loam soil with plenty of organic matter are ideal for starting nursery beds. In case of heavy soil it is desirable to add 2-3 kg sand per square meter so that the seed emergence may not be damage. Before sowing the seeds it is essential to sterilize the bed by soil solarization or formalin or fungicides like captan or thiram@ 4-5 g /m^2. The seeds should be treated with captan or thiram @ 3g/kg seeds.

Beds of 100cm wide and 3-5 m in length and 10-15cm high are ideal. It is desirable to allow 30cm space between two beds. Applications of 5 kg F.Y.M. 200-250g single super phosphate and 200-250g murate of potash per square meter proves beneficial. Manure and fertilizers should be spread uniformly in the beds and mixed. If beds are dry, irrigation of beds helps in proper decomposition of manure and fertilizers. Seed showed be sown 5-7 days after the application of manures and fertilizers.

Sowing of seed and seed bedcover

Seed sowing should be done either by broadcast method or line sowing method. Line sowing facilitate less incidence of damping off disease resulting to reduce seedslings mortality. Several advantages i.e. each and every seeds lings will be healthy, bold and uniform, less seeds are required compared to the broadcast method. Seed bed cover required for better emergence. Therefore, a mixture of sand: soil: FYM in the ratio of 1:1:1 is prepared. Apply 3-4 g thiram or captan per kg mixture if, it is not treated. To maintain the soil moisture for seed germinaton cover the seed bed with a thin layer of mulch of paddy straw sugarcane trash, sarkanda or any organic mulch during hot weather and by plastic mulch in cool weather. It maintains the soil moisture and temperature for better seed germination, suppresses the weeds, protects form direct sunlight and raindrops and against bird damage.

After three days observe the seed beds daily. As and when the white thread like structure is seen above the ground, remove the mulch carefully to avoid any damage to emerging plumules. Always remove mulch in the evening hours to avoid harmful effect of bright sun on newly emerging seedlings.

Aftercare

Irrigate beds with 0.5% urea or 1.0% CAN solution at an interval of 15-20 days, during the early stage of seedlings accelerate the seedling growth. Bavistin or calxin at 0.1% alone or mixed with CAN or urea spray, as it helps in overcoming the problems of damping off disease. One or two appication of Rogor or malathion at 0.01% is necessary to prevent attack of pests.

Hardening

Subjecting plants to adverse conditions to hasten tissue maturation for increasing hardiness, mainly done by withholding irrigation. In this process seedling are given some artificial shocks atleast 7-10 days before uprooting and transplanting. Seedings are exposed to the full sunlight, all the shedding nets, polythene sheets should be removed and irrigation is stopped slowly and slowly. The hardening is done by holding the watering to the plant by 4-5 days before transplanting, lowering the temperature also retards the growth and adds to the hardening processes and by application of 4000 ppm NaCI with irrigation water or by spraying 2000ppm of cycocel.

Transplanting

Too old or too young seedlings both are unsuitable for transplanting in the field. Seedlings having 4-6 true leaf stage or 10-15 cm tall are ideal for transplanting. Seedling can be transplanted bare root or with a soil ball containing roots. Always transplanting is done in the evening and soon after the field is irrigated.

27

Nursery Bed for Seedling Raising

Purpose- One of the important operation for horticulture crops is raising of seedlings (vegetable, fruit trees, shrubs and flowering annuals). Seedlings are either directly planted as in the case of some vegetables, flowering annuals, shrubs and trees or may used as root stock as in the case of mango, citrus etc. For healthy and good stand of seedlings proper preparation of nursery beds is essential. Since large number of seedlings are raised in a consideration. The protection of seedlings from pest, diseases, strong sun and wind is very important. These factor will vary if seedlings are raised under Low Tunnel Polyhouse.

Material and equipment- Spade, khurpi, rake, chain, watering can rose, plastic sheet (250 guage), fungicide and insecticide.

Procedure - Select a sunny situation for preparing nursery bed. Mark out the area required for nursery. Dig the soil about 25-30 cm deep with the help of a spade or hand hoe. Remove weeds, weed roots and stones. Level the land with the help of a rake. Spread dry leaves or paddy straw in a thin layer and burn it. This helps in minimizing the diseases and pests problem. If the soil is heavy clay or silt then mix coarse sand at the rate 3-4 kg per metre sq. After this spread well decomposed farm yards manure (FYM) at the rate of 5-6 kg per meter sq., ammonium sulphate 250g and superphosphate 250g /m^2, mix all these well and leave the land for few days.

Preparation of beds -Nursery beds of 120 cm wide and 10 m in length are considered ideal. In between two parallel beds a space of 25-30 cm should left for cultural operations. The bed should 12-15 cm high during rainy season, while 3.0-5.0 cm during winter season high from the soil surface. Sterilization of nursery beds, before seed sowing reduces the risk of soil-borne diseases. Soil sterilization can be achieved by adopting any one of the following methods. small area, need of moisture and nutrients needs to should be taken into

Physical Methods

(i) **Heat** - Spreading a layer of dry leaves (2-4 cm thick) on the nursery beds area, followed by burning the leaves. The leaf heat will sterilize the soil up-to 3-4 cm depth.

(ii) **Solarization (Solar Heat)-** It is a simple and effective method for soil-sterilization. A black polythene of 200-300 gauge be spread over nursery bed area. The edges of the polythene sheet be pressed in the soil to minimize the air circulation. The beds are to be covered for a period of 4-6 weeks depending on the intensity of sunshine. The beds should made moist before covering with the polythene sheet.

Chemical Treatment

(i) **Formaldehyde (Formalin)** - Moist nursery beds area sprayed with formaldehyde (40%) at rate of 2.5-2. 75 ml per litre water. After spraying the beds should be covered with clear or black polythene. The edges of the polythene need to be sealed with wet soil to make air tight. Polythene should removed only after 10-12 days. Seeds should be sown only after 4-6 days after removing the polythene.

(ii) **Bavistin-**Prepare bavistin solution of 0.1-0.2% sprays the moist beds with bavistin solution. Leave the beds for 2-3 days before use.

(iii) **Trichoderma viridi-** For the treatment of beds use 1.0% solution of trichoderma. Leave the beds for 2-3 days after treatment.

Sowing of seeds - In well prepared beds the seed be sown in miniature furrows. The depth of furrows will depend upon the seed size. For bolder seeds depth of furrow should be kept 5-6 cm and for smaller and fine seed depth should be 1-3cm. The furrows should be covered with a mixture of leaf mould and sand (3:1). Seed treatment before sowing the seeds in the sterilized bed, seed should be treated with suitable chemicals to protect them against soil borne pathogen.

Irrigation- Water should be applied with a watering can fitted with rose. Avoid over irrigation. The frequency of irrigation will depends upon the soil type and air temperature

Care of seedlings- The protection should be against strong sun and rain. Covering the beds with agro-net/mosqutonent at height of a meter will protected the seedlings against strong sunshine and wind. Spraying with fungicide and insecticide at an interval of 15-20 day proves very useful to the seedling health.

28

Use of Plant Growth Regulators in Vegetable

Plant growth regulators are organic compounds occurring naturally in plants as well as synthetic, other than nutrients, which in small amounts promote, inhibit or modify any physiological process in plants.

Basically, PGRs are of two types

(1) **Growth promoter-**Auxins, Gibberellins and Cytokinin

(2) **Growth inhibiters-** Abscisic Acid and Ethylene.

1. Auxin

i. Promote cell elongation.

ii. Responsible for phototropism which means there is differential growth response of the plant to light.

iii. Responsible for geotropism therefore stem shows negative geotropism and roots exhibit positive geotropism.

iv. Cause apical dominance and on removal of apical buds, the lateral buds began to sprout.

v. Help in root initiation.

vi. Help in the development of fruits in the absence of pollination called parthenocarpic.

vii. Control abscission of plant organs.

viii. Help in seed germination.

ix. Help in breaking the seed dormancy.

x. Help in flowering and day neutral plants auxins play as a florigen.

xi. Auxins are selective weed killers.

Natural

Harmon's	Vegetable crops	Affects
IAA	Spinach, Fenugreek, Okra, Tomato, Brinjal, Cowpea, Onion, Cabbage and Cauliflower	Seed germination, Fruit set and yield

Synthetic

IBA, 2,4-D and PCPA	Tomato, Cabbage and Cauliflower	Fruit set, fruiting and yield
2, 4, 5-T	Potato	Seedling set, growth and yield improve shelf life.

2. Gibberellins

1. Able to overcome genetic dwarfism.
2. Maintain balances between internodes growth and leaf development.
3. Cause development of parthenocarpic fruit.
4. Responsible for mobilization of storage compounds durjng germination.
5. Substitute cold treatment for breaking dormancy and cause flowering.
6. Substitute long day requirement of plants and help long day plants to flower in short days. Gibberellic acid (GA) potato-Breaking dormancy Tomato, Cucurbits-Sex expression, fruiting and yield.

3. Cytokinin

1. Stimulate cell division and enhance DNA and RNA synthesis.
2. Stimulate root initiation and growth.
3. Stimulate shoot initiation and growth.
4. Break dormancy of lateral buds due to apical dominance.
5. Retard senescence.
6. Inhibit protein degredation.
7. Help in translocation of assimilates.
8. Influence leaf shape and pigmentation.

Zeatin-Frenchbean, Pumpkin, Artichoke Cell enlargernent.

Kinetin-Vegetables Improve fruit setting,

Delayed senescence.

BA-Lettuce-Break Seed dormancy tissues culture

4. Ethylene

1. Promote ripening of fruits.
2. Responsible for induction of respiratory climacteric
3. Inhibits elongation of stem and root.
4. Stimulates growing cells to grow isodiametrically than longitudinally.
5. Stimulates seed germination in some species.
6. Regulates feedback inhibition mechanism of auxins
7. Induce flowering.

Cucurbits-Flowering, fruiting and sex expression.

Okra-Vegetative growth and apical dominance,

Tomato, chillies-Earliness, fruit ripaning and yield.

5. Abscisic Acid (ABA)

1. Retard growth
2. Induces bud dormancy.
3. Promotes senescence.
4. Promote abscission
5. Initiates flowering in some short day plant in long days.
6. Helps in chlorophyll synthesis.
7. Helps in stomatal regulation.

Synthetic growth regulators

Morphactins

1. Chlorofluocrecol methyl-Encourages lateral branching, inhibit
2. 2,3,5-Triodobenzoic acid-Apical dominance, prevents lodgening, promote flowering

Growth retardants

1. Chlormaquat or Cycocel (CCC)-Retard plant growth,
2. Chlorphonium Chloride or Phosphon-D Delays post harvest senescence of green vegetable
3. AMO- 1618-Productions of healthier vegetable seedlings, reduce
4. Chloropham (CIPC)-Lodgening, anti transpirant
5. B995 (Aminozide)
6. Paclobutrazol (PP333 or Cultar)

Cell division inhibitors

1. Chloropham-Prevention of sprouting, suppression of growth, antiripener
2. Propionate-Growth retardants, axillary bud controller
3. Malcic Hydrazide

Gameticides

2,3-dichloro isobutyric acid-Induction of male sterility

29

Restricted Pesticides in India

Restricted pesticides for use

1. Aluminium phosphide
2 Captafol
3. Carbaryl
4. Dieldrin
5. Methyl bromide
6. Ethylene dibromide (EDB)
7. Sodium cyanide
8. Lindane
9. Methyl parathion
10. Monocrotophos
11. Chlorbenzilate

Pesticides banned for use in agriculture in India

1. Dibromochloropropane
2. Dibromochloropropane (DBCP)
3. Pentachloronitro benzene (PCNB)
4. Pentachlorophenol (PCP)
5. Toxaphene
6. Ethyl parathion
7. Chlordane
8. Heptachlor

9. Aldrin
10. Paraquat di-methyl sulphate
11. Nitrofen
12. Nicotene sulphate
13. Phenyl mercury acetate
14. Tetradifern
15. Calcium cyanide
16. Copper Acotcarsenite
17. Ethyl mercury chloride
18. Menazon
19. Sodium methane arsonate (SMA)
20. Benzene hexachloride (BHC)
21. Phelyl mercury acetate (PMA)
22. DDT
23. Chlorobenzilate
24. Endrin
25. Methomyl
26. Phosphamidon

30

Test Weight of Vegetable Seeds

Purpose: This test is conducted to determine the boldness and soundness of the seeds. The weight of a given number of seeds of different varieties or kinds will vary. This variation is due to boldness and soundness of the seeds. The seeds will vary according to the soil fertility, variety and other environmental conditions. Higher the weight of a variety per thousand seeds, better is the variety for cultivation.

Materials and equipments: Scale, balance and seeds.

Procedure

Draw the sample of the seeds uniformly. Mix them thoroughly. Count 3 to 4 lots (replication) each of hundred seeds for each variety. Record the weight of each lot separately as outlined in observation.

Result

Observation

Date	Replication	Variety Wt. (g)				
		A	B	C	D	Mean
	1234 Total Mean					

Conclusion

Seed sample having bold seed always record higher weight per thousand seed.

Assignment

Determine the thousand grain weight of different variety of okra, field pea and radish.

30

Test Weight of Vegetable Seeds

Purpose: This test is conducted to determine the boldness and soundness of the seeds. The weight of a given number of seeds of different varieties or kinds will vary. This variation is due to boldness and soundness of the seeds. The seeds will vary according to the soil fertility, variety and other environmental conditions. Higher the weight of a variety per thousand seeds, better is the variety for cultivation.

Materials and equipments: Scale, balance and seeds.

Procedure

Draw the sample of the seeds uniformly. Mix them thoroughly. Count four replications each of hundred seeds for each variety. Record the weight of each lot separately as outlined in observation.

Result

Observation

Date	Replication	Variety Weight (g)				
		A	B	C	D	Mean
	Total					
	Mean					

Conclusion

Seed sample having bold seed always record higher weight per thousand seed.

Assignment

Determine the thousand grain weight of different variety of [illegible], cow pea, chick pea and [illegible].

31

Mode of Reproduction

The different modes of reproduction applied in vegetable crops are

Asexual

In this case the vegetative parts of the plants viz. rhizomes, corms, tubers, bulbs etc whole or part there off are utilized to produce new individuals. The typical examples are potato, sweetpotato, parwal, artichoke, colocasia, zinger, turmeric elephant foot yam, etc. There are some vegetative crops which can be reproduced through bcth vegetative part as well as seed e.g. asparagus.

Sexual

In this gametic fusion occur to produce new individuals. The plants may be

i. pre-dominantly self-pollinated

ii. pre- dominantly cross-pollinated

iii. often cross-pollinated.

The vegetable crop may be self- pollinated but some cross-pollination may take place due to the orientation of the flower on the plant and flower morphology

Self Pollination

The transfer of pollen from anther to stigma of the same plant are same flowers

Mechanisms promoting self-pollination

1. Cleistogamy

In this case, flowers do not open at all. This ensures complete self-pollination since foreign pollen cannot reach the stigma of a closed flower eg: lettuce

2. Chasmogamy

In some species, the flowers open, but only after pollination has taken place. In tomato, brinjal, the stigmas are closely surrounded by anthers. Polination generally occurs after the flowers open. But the position of anthers in relation to stigma ensure self pollination.

- In some species, flowers open but the stamens and the stigma are hidden by other floral organs. In several legumes eg. : pea (*Pisum sativum*)
- In a few species stigmas become receptive and elongate through the staminal columns. This ensures predominant self pollination.

CROSS POLLINATION

Pollen grains from flowers of one plant pollinate the flowers of other plant. The transfer of pollen from a flower to the stigmas of the others may be brought about wind (anemophilly) water (hydrophily) or insects (entomophily).

Mechanism promoting cross pollination

There are several mechanism that facilitate cross pollination.

1. Dicliny or Unisexuality

It is a condition in which the flowers are either staminate (male) or pistillate (female)

1a Monoecy

Staminate and pistillate flowers occur in the same plant, either in the same infiorescence or in separate infioresoence eg: cucurbit, sweet corn.

1b Dioecy

The male and female flowers are present on dfferent plants. i.e the plants in such species are ether male or female eg: asparagus, spinach, pointed gourd.

2. Dichogamy

Stamens and pistils of hermaphrodite flowers may mature at different times, thereby, facilitating cross-polination.

a. Protagyny

Maturation of stigma ahead of anther eg: *Brassica species*

b. Protandry

Maturation of anthers ahead of stigma eg: carrot and onion

3. Heterostyly condition

Self pallination is impossible because of relative position of the stigma and anther

a. Dimorphic condition

Flowers with long style and short filament of the anther eg: tomato

b. Heteromorphic condition

Flowers with short style and long anther filament eg: brinjal

A combination of two or more of the above mechanism may occur in some species. This imporves the efficiency of the system in promoting cross-pollination.

4. Salf Incompatibility

It refers to the failure of pcllen from a flower to fertilize the same flower or cther flower on th same plant. Self incompatibility is of two types sporophytic and gametophytic. In both the case-flowers do not set seed on selfing. eg: cauliflower, radish.

5. Male sterility

Male sterility refers to the absence of functional pollen grains in otherwise hermaphrodite flowers

Often cross pollination

In many crcp plants, cross pollination is 5% and may reach 12% such species are generally known as often cross pollinated crop. The architecture of such crops in intermediate between those of self pollinated and crcss pollinated species. Consequently in such species breeding methods suitable fcr both of them may te profitably applied. But cften hybrid varieties are superior to others eg: okra,brinjal, chilli, lima bean etc.

3. Heterostyly condition

Self pollination is impossible because of relative position of the stigma and anther.

a. Dimorphic condition

Flowers with long style and short filament of the anther eg. Jhumka.

b. Heteromorphic condition

Flowers with short style and long anther filament eg. brinjal

A combination of two or more of the above mechanism may occur in some species. This improves the efficiency of the system in promoting cross pollination

4. Self incompatibility

It refers to the failure of pollen from a flower to fertilize the same flower or other flower on the same plant. Self incompatibility is of two types sporophytic and gametophytic. In both the case flowers do not set seed on selfing. eg. cauliflower, radish.

5. Male sterility

Male sterility refers to the absence of functional pollen grains in otherwise hermaphrodite flowers.

Often cross pollination

In many crop plants, cross pollination is 5% and may reach 12%, such species are generally known as often cross pollinated crop. The architecture of such crops is intermediate between those of self pollinated and cross pollinated species. Consequently in such species, breeding methods suitable for both of them may be profitably applied. But hybrids and varieties are superior to others eg. okra, brinjal, chilli, lima bean etc.

32

Hybrid Seed Production of Vegetable

Improved production technology is an integral part of hybrid seed production technology, which is a very high venture practice. This is because like several factors (climate, availability of parental lines and technical labour etc), success of hybrid seed production also depends on the success of crop management in the hybrid seed production field

Important operations

1. Selection and multiplication of parental lines.
2. Isolation distance.
3. Seed bed preparation and sowing.
4. Selection of hybrid seed production field.
    a. Isolation distance
    b. Preparation of field.
5. Transplanting method.
6. Irrigation and earthing up.
7. Crop protection measures
8. Harvesting of F_1 hybrid seed.
9. Extraction and packaging of seeds.

In production of hybrids breeder performs the following sequence operations.

1. Material selection (Needed)
    i. Forcep
    ii. Needle
    iii. Petridish
    iv. Cotton plug

v. Tags

vi. Covering bags.

2. Selection of perental lines.
3. Selection of flowers (flower buds)
4. Emasculation
5. Pollination
6. Tagging
7. Covering the flower after pollination
8. Observation of fertilized flowers
9. Removal and non- crossed flowers
10. Harvesting of crossed fruits
11. Extraction of seeds
12. Packaging of seeds.

33

NPK Content in Manures

A. Concentrated Organic Manures

S.N.	Name of Organic Manures	N%	P_2O_5%	K_2O%
1.	Caster Cake	4.3	1.8	1.3
2.	Karanj Cake	3.9	0.9	1.2
3.	Mahua Cake	2.5	0.8	1.8
4.	Neem Cake	5.2	1.0	1.4
5.	Coconut Cake	3.0	1.9	1.8
6.	Groundnut Cake	7.3	1.5	1.3
Bulky Organic Manures				
1.	F.Y.M.	0.5	0.2	0.5
2.	Farm tiller compost	0.5	0.15	0.5
3.	Night soil	5.5	4.0	2.0
4.	Town compost	1.5	1.0	1.5
5.	Sludes	1.5-3.5	1.5-4.0	0.3-0.6

B. Micro Nutrients and their Composition

S.N.	Micro-nutrient	Farm used of chemical formula	Percentage of elements
1.	Iron (Fe)	Ferrous sulphate ($FeSO_47H_2O$)	20.0
2.	Copper (Cu)	Coper Sulphate ($CuSO_45H_2O$)	25-35
3.	Zinc (Zn)	Zinc Sulphate ($ZnSO_47H_2O$)	22-35
4.	Manganese(Mn)	Manganese sulphate($MnSO_44H_2O$)	23
5.	Boron (B)	Borax or Sodium Borate ($NaBO_710H_2O$)	10.6
6.	Molybdenum (Mo)	Sodium Molybdate($Na_2MO_22H_2O$)	37-39
7.	Calcium (Ca)	Calcium nitrate ($CaNO_3$)	19.5
8.	CAN	Calcium ammonium nitrate	8.1
9.	Chlorine	Ammonium chloride (NH_4Cl)	69-70
10.	Molybdate	Ammonium molybdate (NH_4Mo)	54

34

Numerical on NPK Required

Problem- Calculate the amount of farm yard manure, urea, single super phosphate and murate of potash for one hectare tomato crop which require 100, 60, 60 kg ha N, P, K respectively. 1/3 quantity of nitrogen should be given through farm yard manure and rest through chemical fertilizers

Solution

Total requirement of nitrogen = 100 kg/ha

Phosphorus= 60 kg /ha

Potash = 60 kg/ha

1/3 quantity of nitrogen 100/3= 33.3 kg (should be applied through FYM)

we find 0.5 kg nitrogen From = 100 kg FYM

so that 1.0 kg nitrogen from FYM = 100/0.5= 200kg

Hence 33.3 kg nitrogen from FYM 100/0 5 x33.3kg =6660 kg

These 6660 kg FYM also having Phosphorus and potash, so that it can be calculated.

100 kg FYM provides phosphorus = 0.25/Kg

1kg = 0.25/100

so that 6660 kg provides the phosphorus = 0.25/100 x 6660

= 16.65 kg

100 kg FYM provide the potash = 0.5 kg

1 kg FYM provide the potash = 0.5/100

so that 6660 kg provide the potash = 0.5/100x6660

= 33.31kg

Rest amount of nutrients

Nitrogen = 100-33.3kg = 66.7 kg

phosphorus = 60-16.65kg = 43.35 kg

potash = 60-33.3kg = 26.7 kg

Rest amount of nutrient should be given through chemical fertilizers

So that for nitrogen we calculate the quantity of Urea

46kg nitrogen find from = 100kg urea

So that 1 kg nitrogen find from 100/46

Hence 66.7 kg nitrogen find from urea =(100/46) x 66.7 = 145kg.

For phosphorus we calculate the quantity of single super phosphate

16kg phosphorus find from = 100kg sirgle super phosphate

So that 1kg phosphorus find from= 100/ 16

Hence 43.35 kg phosphorus find from = (100/16)x43.35

= 270.93kg 271 kg

For potash we calculate the quantity of murate of potash

60kg potash find from= 100kg murate of potash

1kg potash find from= 100/60kg murate of potash

So that 26.7kg potash find from = (100/60)x26.7

= 44.5 kg

Result

Farm yard manure	= 6660kg
Urea	= 145kg
Single super phosphate	= 271 kg
Murate of potash	= 44.5 kg

35

Nurseries and Seed Stores of Uttar Pradesh

Important nurseries and seed merchants of Uttar Pradesh

1. Azad Nurseries, Laxmi Gate, Saharanpur.
2. American Seed Store, Ayodhya.
3. Ashoka Seed Farm (Regd.) Darwaja Road, Ayodhya
4. Beiz Nursery, Azmatgarh Palace, Varanasi.
5. Ganesh Bagh Nursery, Kamachchha, Varanasi.
6. Ram Das & Co. Kamachchha, Varanasi.
7. Manjari Paudhshala, Line Bajar, Jaunpur.
8. Green India, Jagdispur, Jaunpur.
9. Gupta Nursery, Pilkhni Gaurabadsahpur, Jaunpur.
10. Vanaspati Beej Bhandar, Jaunpur.
11. Suraj Beej Bhandar, Kotawali, Jaunpur.
12. Ramchandra Bhageluram, Old Sabji Mandi, Jaunpur.
13. Gulab Singh Kalian Singh, Motibagh, Jamuna Bridge, Agra.
14. Jagdamba Nursery, Juhibagh, Agra.
15. L.R.Brother's Seedsman, Saharanpur
16. Manjulikey Garden, Deoband Road, Saharanpur.
17. Mangalsen Ram Prasad, Motibagh, Jamuna Bridge, Agra
18. Manohar Beejalaya, Quilla, Bareily
19. National Batasi Lal Choudhary, Dehrabagh, Jamuna Bridge, Agra

20. Puran Singh Krishna Nursery Jamuna Bridge, Agra.
21. Rodhey Shyam & Co., Seed Merchant, Subji Mandi, Bareily
22. Rama & Co. Seed Merchants, Saharanpur.
23. Safdar Seedsman and Nurseryman, Malihabad, Lucknow
24. S.B. Brother's Seeds man, Kahanibagh, Saharanpur
25. The Henbane Nursery, Saharanpur.
26. Union Nursery, Subjimandi, 1330, Niawan Road, Ayodhya
27. The Indian Seed Co., Seedman 1330, Niawan Road, Ayodhya.
28. U.P. Breej Bhandar, Chipitola, Agra.
29. Verma and Co., Seed Store, Bareilly
30. Verma and Co., Seed Merchant, Kannuaj.
31. Incharge, Vegetable Research Farm, Kalyanpur, Kanpur.
32. Director, Indian Institute of Vegetable Research, Araji Line, Varanasi
33. Hind Seed Company, Ayodhya.
34. Pandey Seeds Pvt. Ltd., Arya Kanya Road, Ayodhya.
35. L.M.L. Seeds Company, 403 Niyawan Road, Ayodhya.
36. The Fyzabad Seed Co., Gajadhar Bhawan, Niyawan Road, Ayodhya

36

Insect Pest of Vegetable Crops

Important insect-pests of vegetable crops

1. Aphids: (*Myzus persicae*)

The aphids damage the crop by sucking the cell sap from the leaves as stems. The affected leaves in severe cases curl-up and rolled. The plant may be stunted, wither and die. There are three different species of aphids. They are greenish or black in colour.

Control

By spraying Rogor 30EC or Metasystox 25EC or Dimecron (Phosphamidon) or Malathion @ 0.02% at 10days intervals.

2. Jassids : (*Empoasca devastans*)

They are green coloured small leaf hopper. They are found mostly on the under side of the leaves. They suck the sap from the leaves causing them to curl-up and drying of the leaves.

Control

Spraying of Rogor or monocratophos 0.02% at 10 day interval are effective.

3. Epilachna beetle (*Epilachna* sp.)

They are small. Oval shaped beetle with a few spots on their back. Both grub and adult feed on the leaves and make them lace-like appearance.

Control

Spraying of Rogor or monocratophos 0.02% at 10 day interval are effective.

4. Tomato fruit worm: (*Heliothis armigera*)

The caterpillar feed on the leaves and finds its way inside the fruits by making holes.

Control

1. Affected fruits should be collected and destroyed.
2. Spraying with endosuphan or diemecron with 0.02 % at weekly intervals.

5. Cutworms: (*Agrotis* sp.)

They damages the young seedlings by cutting them at the base near the soil and finds their way inside the potato tubers by making a hole. They are active during night and causes extensive damage.

Control

Application of Aldrin 30EC or Heptachlor is effective.

6. Brinjal Shoot and Fruit borer: (*Leucinodes orbanlis*)

This is one of the serious pest and a limiting factor for the cultivation of brinjal. The pinkish larvae attack the terminal shoots and bore inside as a result the shoot dry and wither away.

Control

i. A soon as there a withering sign of the shoots it should be removed and destroyed.
ii. Fruits showing and boring sign should be picked up and destroyed.
iii. Spraying of dimecron or endosulphan or malathion@ 0.02 % control the insect

7. Thrips (*Scirtothrips dorsalis*)

These small insects suck the sap from the leaves, which causes curling of leaves and stunting of the plant.

Control

Soil application of Aldicarb or Carbofuran followed by spraying Monocrotophos 0.05 % at 10 days intervals.

8. Diamondback moth (*Phutella xyllostella*)

It is serious pest and active durirg October-November. Larvae feed on the leaves, sometime they make holes on the leaves.

Control

Spraying of Endosulphan at 10 days intervals.

9. Mustard saw fly (*Athalia* sp)

The minute black fly lays egg inside the leaf tissues. The caterpillar feed on the leaves of the young seedlings. The black caterpillar attack all most all the cruciferous vegetables.

Control

Spraying Malathion 0.02% at fortnightly interval is effective.

10. Semi-Looper (*Plusia* sp.)

The green coloured caterpillar feed on the leaves of cauliflower. They can be identified by the characteristic loop they form, while they move. They attack almost all the cruciferous crops.

Control

Spraying Malathion 0.02% at fortnightly interval is effective.

11. Cabbage butterfly: (*Pleris* sp.)

The green coloured caterpillars of the butterflies feed on the leaves from the margin and proceed to the centre and skeletonise them. The young caterpillars feed on the tender shoots, leaves and head of cabbage.

Control

Spraying of Malathion or Diazinon or Parathion at 0.02 % is effective.

12. Onion maggots (*Hylemic antique*)

It is small gray fly smaller then house fly. They lay egg in the soil near the base of the plant and hatch within a week. Maggots enter the bulb through roots and attack the tender portions.

Control

i. Crop rotation.

ii. Use of aldrin or dieldrin is effective.

13. Spotted boll worm: (*Earias* sp)

The larvae bore inside the growing shoot, flowers and fruits and damage the plant and the fruit.

Control: Same that of jassid.

14. Mites: (*Telranychus cinnabarimus*)

The nymphs and the adult suck the sap from the leaves and stem. Leaves become whitish and plant, stunted.

Control

Spraying Metasystox 25EC or Rogor 30EC@ 0.02 %.

15. Nematodes: (*Meloidogyne* spp., *Rotylenchus* spp.)

It attacks the Solanaceous crops. It attacks the roots by producing tiny galls. The attacked plants become stunted, show sign of wilting, leaves starts wilting.

Control

i. Soil fumigation with DD (Dichloropropene-dichloro propene) or EDB Nemagon is effective or Ethyle Dibromide 3g/ plant.

ii. Temik or Nemafos 46 @ 20 Kg/ha as soil application.

16. Termite: (*Odontotermus albus*)

They attack the root, bark and the whole plant. The plant becomes week and stunted.

Control

i. Apply Heptachlor Dust25 Kg/ha.

ii. Apply Furadon 3G. 3-5g/ plant.

iii. Folidal 2% Dust-10-15 g/plant

iv. Methyle Parathion-2 % Dust-10-15g/plant

v. Chloropyrypros 2% Dust-5-10g/plant.

Insecticides Used Against Vegetable Crops

1. Aphids- Dimethoate 30EC, Phosphamidon 85 WSC
2. Thrips-Dimethoate 30EC.
3. Leaf Hopper- Phosphamidon, Methyle demeton
4. Mites - Dicofol 18.5 EC, Tetradifon 20.
5. Scales-Quinalphos 25EC, Fenitrothion, Fenthion.
6. Mealy bug-Fenitrothion, Quinalphos 25EC. Endosulphan 35EC.

Commonly used Insecticides

Insecticides	Mode of action
Aldicarb	Sy,C.
Carbary (Sevin)	C
Carbofuran (Furadon).	Sy,C.S.
Endosulfan (Thiodon)	C.S.
Fenitrothion.	C.S.F.
Fenthion	E.S.
Malathicn	Sy,C.
Cypermethrin	C,S.
Diazinon	C.F.
Dimethoate (Rogor)	Sy.C.F.
Permethrin	Sy,C.
Dicofal	C
Phosphamidon (Dimecron)	Sy,C,F.
Furadon	C
Monocrotophos (Nuvacron)	C,S.
Dichlorvos (Nuvan)	C,F.
Malathion	C,S.
Diazinon	C,F.

C-Contact poison, F-Fumigant, S-Stomach Poison, Sy-Systemic poison

37

Diseases of Vegetable Crops

Important diseases of vegetable crops

1. Damping off: (*Phytophthora parasitica, Pythium sp. Rhizoctomia solani)*:- This is a disease of the seedlings at the nursery stage. Due to the attack of this fungus seedlings rot at the ground level and collaps.

Control

1. Before sowing the seeds should be treated with ceresan, Captan, Thiram or Agrosan GN@ 0.02% or 2g/ 500g of seeds
2. The soil of the nursery bed should be treated with formaldehyde or formation 5 % or brassicol
3. Seedling should be sprayed with Captan or Blitox@ 0.0 %
4. Hot water treatment of seeds at 51.7°C for 30 minutes.

2. Early Blight: (*Alternaria solani, A brassicae and A. brassicola)*:- The fungus attacks the leaves, stem and fruits. They make dark brown spots, usually sunken spots

Control

1. Seeds from infected plants should not be used.
2. Seeds should be treated with copper fungicides before sowing.
3. Spray of Dithane Z-78 is effective.

3. Late Blight: (*Phytophthora infestans)*- The fungus attack the leaves, stems and fruits. Dark water- soaked patches appear at different parts of the plant. There is profuse fluffy growth of the fungus is severe attack.

Control: Spraying of Dithan Z-78 and Copper Oxychloride is effective.

4. Fusarium wilt: (*Fusarium oxysporum, Ozonium* sp. *Or verticillium* sp.):- The lower portion of the leaves are affected, which become yellow between the veins and in course of time wilt and drop off along with the petioles.

Control

1. Healthy seeds should be used
2. Crop rotation should be followed.
3. Resistant varieties should be used.
4. Seed soaking in 0.05% hydroquinine or Bavistin.

5. Bacterial wilt: (*Pseudomonas solanacearum):*- This is a soil born disease. Lower leaves starts wilting, yellowing of foliage following by wilting and collapse of the plant.

Control

1. Crop rotation be followed.
2. Resistant varieties should be grown.
3. Infected plants should be carefully collected and destroyed.

6. Powdery mildew: (*Oidium* sp. *Sphaerotheca fuliginea, Erysiphe* sp.): The vegetative parts are first affected as white powdery, fluffy growth on the under surface of the leaves. In severe cases it spread of to entire vegetative parts of the plant. Plants do not grow properly and fruiting is affected

Control

Spray cf wettable sulphur or kerathane 0.2%.

7. Downy mildew: (*Peronospora parasitica):*- It is prevalent in areas of high humidity. The disease appears as yellow or whitish irregular spots on the upper surface of the leaves. This iş a serious disease of cucurbits.

Control

1. Spraying or dithane M-45 at 10 days intervals at 2 to 3 times.
2. Spray of Difoltan is also effective.

8. Black leg: (*Phoma lingam):*- The fungus attack when there is stagnation of water in the field. The plant at the base of the stem and root are affected and wilt.

Control

1. Crop rotation.
2. Seed treatment
3. Resistant varieties should be grown.

9. Club rot: (*Plasmodiophora brasicae)*:- The roots of the plants show, club like appearance. There is swelling of the roots. The plant recovers during sunny days and recovers towards evening.

Control

1. Crop rotation.
2. Seedling should be treated with mercuric chloride.

10. Anthracnose: (*Collectotricum* sp.)*:* This is the common disease of cucurbits. Reddish brown, dry leaf spot appears. The spots are sunken, dark appears as canker on stem, leaves and pods. The fungus is seed borne.

Control

1. Use of healthy seeds.
2. Seeds should be treated with Thiram or cereson @ 2g/ Kg of seeds.
3. Spraying of Benlate or Dilthane Z-78 or Zineb or Macozeb or Bavistine is effective.
4. Spraying with copper fungicides like Blitox Fytolan@ 0.2 % is effective.

11. Leaf Spot: (*Cercospora cruenta)*: This is a common disease of the cucurbits. The lower leaves are first affected water soaked lesions appears on the leaf lamina and also on the pods.

Control: 1 Spray Copper Oxychloride 0.2% or 2g/l.

12. Leaf curls Virus:- The disease is characterized by curling of leaves, uneven leaf surface, small leaf size and stunted plant growth. This disease is spread by white fly.

Control

1. White fly should be controlled by spraying Rogor or Dimecron @ 0.02 % at weekly interval,

13. Scale: (*Streptomyces scabies)*- This is a common disease of potato. Affected tubers show dark, circular lesions about 0.1 cm diameter.

Control

1. Disease free tubers should be used.
2. Tubers are treated in 0.5% Agallol solution for 10 minutes
3. Soil should be treated with Brassiccl.

Some used fungicides against plant diseases

1. Benlate - Rots, spots.
2. Captan - Damping off, rots, spots wilt.
3. Carbendazim (Bavistin spcts)- Powdery mildew, rat, spots.
4. Dinocap (Karathane) - Powdery mildew
5. Captafol-Spots
6. Mancozeb (Dithane M-45)- Anthracnose, rot, spots, damping off, mould.
7. Carbofuran (Funadon)-Soil application
8. Streptocycline-Bacterial diseases
9. Thiophanate, Wettable Sulphur-Powdery mildew
10. Zineb (Dilthane Z-78)-Leaf spot, rot, blight
11. Ziram-Anthracnose, leaf spot, blight.
12. Thiram - (Indofil M-45)-Anthracnose, Rots, spots.
13. Carbendazım (Bavistin)-Powdery mildew, leaf spots, root rot, stem rot.
14. Brassicol-Damping off, root rot.
15. Fytolon, Blitox-Die back, Anthracnose, spots.
16. Furadon, Phorate-Termite.

Section C
Floriculture and Landscaping

38*

Flowering Trees

*Table starts from next page.

Description of the important flowering trees

S.N.	Common name	Botanical name	Family	Flower colour	Flowering time	Method of Propagation	Remark
1	Amaltas the java cassia	*Cassia fistula* *Cassia javanica* *Cassia nodosa* *Cassia siamea*	Leguminosae	Golden Yellow Pink Pink Yellow	April-May May- June May- June May- June	Seeds	Medium sized 10 tall Beautiful flower Medium sized tree Medium sized beautiful Flower avenue tree
2	Gulmohar	*Poinciana regia syn. Delonix regia*	Leguminosae	Orange/Scarlet	April-May	Seed	Flower cover the whole tree
3	Pride of india	*Lagerstroema flosreginae*	Lythraceae	Purple mauve	April-May	Seeds	Medium size tree
4	Blue gulmohar	*Jacaranda mimosifolia*	Bigoniacea	Blue mauve	March-May	Seed/semi hard wood cutting	Elegant,deciduous tree
5	Bottle brush	*Callistemon lanceolatus*	Myrtaceae	Scarlet	March-August September	Seeds layers	Smalldrooping branches flower born in spikes
6	Flame of forest	*Butea monosperma* *B. monosrerma var. tutea*	Leguminosae	Scarlet orange yellow	April-May	Seed	Medium size, rough trunkdeciduous
7	Kachnar	*Bauhinia purpurea* *B. alba*	Leguminosae	Purple deep pink white	November	Seed	Medium size
8	Champa	*Michelia champaca*	Magnoliaceae	White/ creamy yellow	April-May/ Sep-Oct	Seed/grafting	Medium size long leaves
9	Pangri	*Erythrina indica*	Leguminosae	Scarlet red	Feb-April	Cutting	Fast growing, densly branch
10	Madar tree	*Eliricidia maculate*	Leguminosae	Pale Pink	Feb-March	Seed	Good looking, medium used for shading cocoa
11	Amherstia	*Amherstia nobilis*	Leguminosae	Vermilion yellow tip	Feb-May	Seed/ Leyering	Medium, beautiful tree
12	Indian lilac	*Logerstroemia indica*	Lythraceae	Pink maure, white	May-Aug	Seed	Swall tree
13	Him champa	*Magnolia grandiflora*	Magnoliaceae	Pink white	April-May	Layering	Evergreen tree

Contd.

14	Harsingar/seuli	*Nyctanthes arbor-tristis*	Oleaceae	White with orange red tube	Sept- Nov	Seed/ cutting	Sweet scented flower open at night & fall at day tree
15	Akashneem	*Millingtonia hortensis*	Bignoniaceae	Silvery white		Seed/ root suckers	Quick growing straight 20 mtr
16	Yellow gulmohar	*Peltophorum inerme*	Leguminosae	Yellow	Feb-May/ Sep-Aug	Seed/ cutting	Beautiful ornamented tree
17	Temple tree/ pagoda tree	*Plumeria alba*	Apocynaceae	White	March-April/ July-Aug	Cutting	Evergreen tree
18	Chameli, Gul-e-chin	*Plumeria rubra*	Apocynaceae	White-rose	Oct-Dec.	Cutting	Low growing 6-8m tall tree
19	Sita ashoka	*Saraca indica*	Leguminosae	Orange	April-June	Seed	8-10m tall evergreen tree
20	Fountain tree	*Spathodea campanulate*	Bignoniaceae	Scarlet ofange/ crimson	Feb- March	Seed /root-suckers	20-25m tall ornamental
21	Tecoma	*Tecoma argentea*	Bignoniaceae	Bright yellow	Feb- March	Seed	5-8m tall tree

39*

Foliage Trees

*Table starts from next page.

Description of the important foliage trees

S.N.	Common name	Botanical name	Family	Flower colour	Flowering time	Method of propagation	Remark
1	Neem	*Azadirachta indica*	Meliaceae	Small white	April	Seed	12-18m tall tree
2	Christmas tree	*Arucaria cookie*	Pinaceae	-	-	Seed	Ornamental tree
3	Siris	*Albizzia lebbek*	Leguminosae	-	-	Seed	Fast growing spreading tree
4	Jhau	*Casuarinar equisetifolia*	Casuarinaceae			Seed	10-20m tall tree
5	Shisam	*Dalbergia sisso*	Leguminosae			Seeds/ cutting	15-20m tall tree
6	Banyan	*Ficus benghalensis*	Moraceae	Crimson berries	Aug-Sep	Seeds/ cutting	15-20m tall tree
7	Pakur	*Ficus infectoria*	Moraceae			Seeds/ cutting	15-20m tall evergreen spreading tree
8	Pipal	*Ficus religiosa*	Moraceae			Seed	20-25m tall huge tree
9	Indian rubber plant	*Ficus elastica*	Moraceae			Air layering	10-12m quick growing tree
10	Silver oak	*Grevillea robusta*	Proteaceae	Golden yellow	April-May	Seed	15-20 evergreen tree
11	Mahua	*Madhuca indica*	Sapotaceae	Creamish white	March-April	Seed	15-20 tall deciduous tree
12	Mulsari	*Mimusops elengi*	Sapotaceae	White	April-July/ Sept-Nov	Seed	10-15m tall dense evergreen tree
13	Ashoka	*Polyalthia longifolia*	Anotaceae			Seed	10-15m tall dense evergreen tree
14	Chir	*Pinus longiflolia*	Pinaceae			Seed	15-20m tall needle like leaves
15	Putranjiva	*Putranjiva roxburghii*	Euphorbiaceae			Seed	10-12 m tall dense evergreen tree
16	Poplar	*Populus deltrolides*	Saliceaceae			Cutting	Fast growing, deciduous 5-10m tall
17	Karanj	*Pongamia glabra*	Leguminosae	Lilac	April-May	Seed/cutting	5-10m deciduous tree
18	Arjun	*Terminalia arjuna*	Combretaceae	Yellowish white	March-June	Seed	15-20m tall evergreen

Contd.

							avenue tree
19	Morpankhi	*Thuja occidentalis*	Pinaceae			Seeds	5-8m tall, foliage is fern like
20	Travellers tree	*Ravenala madagascariensis*	Musaceae			Seed	3-4m tall, banana like leaf
21	Sterculia	*Sterculia alata*	Sterculiaceae			Seed	8-10m tall, leaves are larg, avenue tree
22	Juniperus	*Juniperus chinenss*	Pinaceae			Seed	5-10m tall, hardy, dense, pyramidal decorative tree
23	Chalta	*Dillenia indica*	Dilleniaceae	Large white	July	Seed/ stem cutting	8-20m tall, slow growing evergreen tree
24	Chitwan	*Alstonia scholaris*	Apocynaceae	White- greenish	April-May	Seed	Strong small during night, 6-10m tall tree

40*

Flowering Shrubs

*Table starts from next page.

Description of the important flowering shrubs

S.N.	Common name	Botanical name	Family	Flower colour	Flowering time	Method of propagation	Remark
1	Gandharaj	*Gardenia jasminoides syn. G. florida*	Rubiaceae	White	April-June	Air-layering cutting	Popular fragrant 3m tall
2	Gurhal	*Hibiscus rosa-sinensis*	Malvaceae	Rose-scarlet		Air-layering cutting	Bushy, 1-2m tall
3	Thalkamal	*Hibiscus mutabilis*	Malvaceae	White in the early morning and canges to pink and deep rose as the day advances	Throughout the year	Seeds cutting	3m tall shrub
4	Rukmini/Raktak	*Ixora coccinea* *I singaporensis*	Rubiaceae	White Scarlet Scarlet	April-June/ July-Sept, Throughtout the year, Greater parts of the year	Cutting/ Layering	2m tall bushy
5	Musaenda	*Musaendra frondosa*	Rubiaceae	White	April-Sept	Layering	Beautiful shrub
		Musaendra luteola		Yellow	April-Sept	Layering	Beautiful shrub
		Musaendra philippica		Pink	April-Sept	Layering	Beautiful shrub
6	Kamini	*Murraya exotica*	Rubiaceae	White		Seed/Layering	Used for hedge topiary
7	Glaucous	*Cassia glauca*	Leguminosae	Yellow	Throughout the year	Sed	Beautiful shrub
8	Karonda	*Carissa carandas*	Apocynaceae	White		Seed	Used for hedge, berries
9	Krishnachura	*Caesalpinia pulcherrima*	Leguminosae	Orange-scarlet	Throughout the year	Seed	Beautiful shrub
10	Radhandra	*Caesalpinia pulcherrima var. flava*	Leguminosae	Yellow	April-Aug	Seed	Beautiful shrub

Contd.

11	Calliandra	*Calliandra inaequilatera*	Leguminosae	Red, pink white		Seed/ layering	Power puff like flower
12	Bougainvillea	*Bougainvillea spp.*	Nycataginaceae	Various colour	Feb-June/ Sept-Dec	Cutting	Versatile shrub
13	Allamanda	*Allamanda cathartica*	Apocyraceae	Yellow	April-Aug	Cutting/ layering	Can be used as climber
14	Clerodendron	*Clerodendron inerme*	Verbenaceae	White	July-Aug	Cutting	Used for hedge
15	Raat-ki-rani	*Cestrum nocturnum*	Solanaceae	Creamy yellow	April-July	Seed/cutting	Fragrant during night bushy, quick growing
16	Day- Queen	*Cestrum diurnum*	Solanaceae	White	Summer	Seed/ cutting	Bloom during day time
17	Orange cestrum	*Cestrum auantiacum*	Solanaceae	Orange, Yellow	Feb-March	Seed/ cutting	Bloom flower
18	Crossandra	*Crossandra* spp.	Acanthaceae	Yellow orange, brick red	Feb-June	Seed/ cutting	Flower are used for making gujras
19	Golden dew drop balchari	*Duranta plumieri*	Verbenaceae	Blue		Seed/ cutting	Used for making hedge
20	Hamelia	*Hamelia patens*	Rubiaceae	Orange red	April-Aug	Cutting, layering	Used for making good hedge
21	Barbadose cherry	*Malpighia glabra M. emasginata*	Malpighiaeae	Purple	April-Aug	Cutting seed	Specimenplant, cherry like fruit
22	Kanel	*Nerium oleander*	Apocynaceae	White, pink, red	April	Cutting layering	Popular shrub
23	Chitra	*Plumbago capensis syn. P. auriculata*	Plumbaginaceae	Blue	Feb-Sep	Cutting layering	Used for edging
24	Poinsettia	*Poinsettia pulcherrima*	Euphorbiaceae	Red	Dec- Feb	Cutting	Rapid growing sun loving
25	Tulsi	*Rauwolfia chinensis*	Apocynaceae	White		Seeds	Evergreen medicinal shrub
26	Coral pant	*Russelia juncea*	Scrophulariaceae	Coral red	March-Aug	Cutting	Pendulous branches
27	Chadni, Taggar	*Tabernaemontana coronaria, T. divaricata*	Apocynaceae	White	March-Aug	Cutting	Used for making hedge
28	Pilakaner	*Thevetia nerifolia*	Apocynaceae	Yellow	Throughout the year	Seeds/ cutting	Funnl shaped flower
29	Bela	*Jasminum sambac*	Oleaceae	White	May- June	Cutting	

41*

Foliage Shrubs

*Table starts from next page.

Description of important foliage shrubs

S.N.	Common name	Botanical name	Family	Method of propagation	Remark
1.	Aglaonema	*Aglaonema* spp.	Araceae	Cutting division	Perennial, leaves green with marking of grey or variegated or silver.
2.	Alocasia	*Alocasia* spp.	Araceae	Cutting tubers, rhizomes	Beautiful, hardy, indoor plant
3.	Aralia	*Aralia* spp.	Araliaceae	Cutting	Hardy lant
4.	Sataber	*Asparagus* spp.	Liliaceae	Seed suckers	Beautiful bristle like cladodes
5.	Colocasia	*Colocasia* spp.	Araceae	Inbers	Attractive foliage
6.	Cordyline	*Cordyline* spp.	Lilaceae	Node cuttings suckers	Related to bracaena
7.	Dracaena	*Dracaena* spp.	Liliaceae	Node cuttings suckers	Long leaves
8.	Cycas	*Cycas* spp.	Cycadaceae	Suckers	Good for pod. greenhouse and indoor plant
9.	Dieffenbachia	*Dieffenbachia* spp.	Araceae	Cutting	Pot plant, warm, humid contions plant are poisonours
10.	Rubber plant	*Ficus elastic*	Moraceae	Air layering, cutting	Pot plants
11.	Maranta (calathea)	*Maranta* spp.	Marantaceae	Suckers, cutting	Good pot plant
12.	Monster	*Monstera acuminate*	Araceae	Division of nodes	Root climbers with good foliage
13.	Philodendron	*Philodendron* spp.	Araceae	Division of nodes	Different shapes of foliage
14.	Pilea	*Pilea muscosa syn.B. microphylla*	Urticaeae	Stem cutting	Grown in shades place
15.	Sansevieria	*Sansevieria cylindrica* *S. trifasciate*	Liliaceae	Division for suckers	Perennial, erect, sword like leaves
16.	Tradescantia	*Tradescantia albiflora*	Commelinaceae	Stem cutting	Low growing, trailing or creeping

42*

Climbers

*Table starts from next page.

Descriprtion of important climbers

S.N.	Common name	Botanical name	Family	Flower colour	Flowering time	Method of propagation	Remark
1	Rangoon creeper (madhu- malati)	*Quisqualis indica*	Combretaceae	White and pinkish	Most part of the year	Cutting layering	Flowers are produced in drooping branching
2	Jhumkolata (passion flower)	*Passiflora edulis, p.alba*	Passifloraceae	White, pink, purple	June-Nov	Suckers layering	Vigorous hardy
3	Madhabilata	*Hiptage beneghalensis*	Malpighiaceae	White with yellow	Dec-Feb	Seeds	Sweet scented evergreen
4	Allamanda	*Allamanda cathartica* *A. Hendersonii*	Apocynaceae	Yellow	April-July	Cutting layers	Easy to grow
5	Coral creeper	*Antigonon leptopus*	Apocynaceae	Orange yellow	Most part of the year	Seed, cutting layering	Tuberous rooted quick growing
6	Duck flower Pelican swan flower	*Aristolochia elegans* *A. Grandiflora*	Polygonaceae	Rose red	April-June	Seed, cutting layering	Flower emit repelling odouck
7	Dutchman's pipe	*Aristolochia denocalymma* *Aristolochia alliaceum*	Bignoniaceae	White purple Brown	March-June	Layering cutting	Emit a garlic like small
8	Trumpet creeper	*Campsis grandiflora* syn. *Tecoma grandiflora*	Bignoniaceae	Yellow pink mauve	July-Aug	Suckers cutting	Trampet shaped flower with aerial rootlet
9	Clerodendron	*Clerodendron splendens*	Bignoniaceae	Red deep orangre	Dec-Feb	Suckers layer	Beautiful climbr
10	Golden shawer	*Bigonia venusta*	Bignoniaceae		Jan-Feb	Layering cutting	Tabular flower grown on compound wall
11	Purple wreath	*Petrea volubilis*	Verbenaceae	Red	Feb-April	Suckers cutting Layering	Star shaped flower
12	Juhi	*Jasminum auriculatum*	Oleaceae	White	April-Sept	Cutting	Sceuted flower
13	Safe bel bridal bouqut	*Porana paniculata*	Convovulaceae	White	Aug-Oct	Cutting, layering seed	Head shaped leaves used for screen
14	Heavenly blue	*Thunbergia grandiflora*	Acanthaceae	Blue with yellow	Feb-Aug	Seed cutting	Used for covering wall
15	Vernornia	*Vernornia elaegnaefolia*	Compositae	White	July-Aug	Seed cutting	Used for screening
16	Money plant	*Pothas aurens*	Areceae	White, Pale Pink		Cutting	

43*

Annuals

*Table starts from next page.

Description of important annuals

A. Winter season annuals (Seeds are sown in Sept.-Oct. and transplanting in Oct.)

S.N.	Common name	Botanical name	Family	Flower colour	Plant height (cm)	Remark
1	Paper flower	*Acroclinum roseum*	Compositae	White and pink	45-60	Daisy-like- flower
2	Flors flower	*Ageratum mexicanum*	Compositae	Blue, white pink	20-45	Fliff hads of flowers
3	Hollyhock	*Althaea rosea*	Malvaceae	Pink, scarlet red, maure vilet and yellow	100-120	Majestic plant
4	Sweet alyssum	*Alyssum maritimum*	Cruciferae	White, pink lilac	10-30	Edeging and bedding purpose
5	Snapdragon	*Antirrhinum majus*	Scrophulariaceae	White, pink, yellow foick red and maroom	30-70	Flowers like dragons jaw on spikes
6	African daisy	*Arctotis grandis*	Compositae	White, orange red	45-60	Bushy, vigorous
7	Daisy	*Bellis perennis*	Compositae	White , pink and crimson	20-30	Dwarf plants
8	Swan river daisy	*Brachycome iberidifolia*	Compositae	White, pink	20-50	Dwarf plant
9	Calendula	*Calendula officinalis*	Compositae	Yellow orange	30-50	Meaning first day of the month
10	Corn flower	*Centaura cyanus*	Compositae	Blue, pink white	60-80	Weed of the corn field
11	Sweet sultan	*Centaura moschata*	Compositae	White, purple, red, yellow	70-10	Flowers look like power/puff
12	Wall flower	*Cheriranthus cheiri*	Crucifereae	Yellow, burnt orange	30-45	Bedding and pot culture
13	Annual Chrysanthemum	*Chrysanthemum coronarium*	Compositae	Yellow white	90-120	Bedding purpose
14	Cineraria	*Cieneraria cruentus*	Compositae	Puple white	30-45	Pot culture, shade loving
15	Clarkia	*Clarkia ebegans*	Onagraceae	White pink, rose scrlate	60-80	Have long spikes
16	Parrot Bill	*Clianthus dampieri*	Leguminosae	Crimson	60-80	Flower are pendulous
17	Coreopsis	*Coreopsis tinctoria*	Compositae	Yellow crimson, brown	45-60	For bedding purposes
18	Cosmos	*Cosmos bipinnatus*	Compositae	Pink, purple, crimson, white	60-100	Good for mass planting
19	Dahlia	*Dahlia variabilis*	Compositae	Yellow, red, blue, white, crismon	60-120	Bedding and pot culture
20	Larkspur	*Delphinium ajacis*	Ranunclaceae	Violat, crimson, pink, blue, white	45-20	Bedding and cut flower
21	Sweet William	*Dianthus barbotus*	Caryophyllaceae	Pink, crimson, mauve, yellow	30-45	Scentd flower
22	Carnation	*Dianthus caryophyllus*	Caryophyllaceae	Pink,white, yellow, mauve, crimson violet	30-60	Cut flower good vaselife

Contd.

23	African daisy	*Dimorphotheca calendulacea*	Compositae	Yellow, white	30-60	Flowers open during day time
24	California poppy	*Eschscholzia californica*	Papaveraceae	Yellow, orange	30-45	Bedding and pot culture
25	Blanket flower	*Gaillardia pulchella*	Compositae	Yellow, orange, lemon, maroom	30-45	Bedding and cutting
26	Freasure flower	*Gazania splendens*	Compositae	Yellow, orange white,red	20-25	Rock garden
27	Boy's breath	*Gypsophila elegans*	Caryophyllaceae	White, pink	30-45	Small lance shaped floor
28	Everlasting flower	*Helichrysum bracteatum*	Compositae	Yellow, pink, red	45-60	Flower long lasting
29	Candytuft	*Iberis amara*	Crucifereae	White lilac	20-30	Flower appear in tufts on spikes
30	Sweep pea	*Lathyrus odoratus*	Leguminosae	White, blue, pink,mauve	90-120	Grown as background,in tufts on spickes
31	Statice	*Limonium sinuatum*	Plumbaginace	White, pink, purple	45-60	Good cut flower
32	Linaria	*Linaria bipartite*	Scrophulariaceae	White, blue, pink, red yellow	30-45	Bed and pot culture
33	Lupinus	*Lupinus lurens*	Leguminosae	Yellow, blue	40-60	Cutting beds borders
34	Stock	*Matthiola incana*	Cruciferae	White, red	60-90	Scented cutting
35	Ice plant	*Mesembryanthemum tricolor*	Aizoaceae	Pink, white, yellow	20-30	Rock garden dry wall
36	Nemesia	*Nemesia strumosa*	Scrophuariaceae	White, blue, yellow, orange	45-60	Cut flower
37	Nigella	*Nigella demascena*	Ranunculaceae	White, blue, rose	45-60	Good colour for cut flower
38	Shirley poppy	*Papaver roheas*	Papaveraceae	Scarlet pink white	60-75	Popular flower
39	Phlox	*Rhlox drummondii*	Polemoniaceae	Pink, crimson, purple, violet	30-45	Bedding plant
40	Nasturtium	*Tropacolum mojus*	Tropacolaceae	Yellow, orange, red	30-40	Bedding
41	Salvia	*Sativa splenders*	Labatae	Red, white, blue, scarlet	30-45	Shade loving
42	Saponaria	*Saponaria calabrica*	Caryophyllaceae	Pink, white, red	60-90	Flowers resembles gypsophilla
43	Pansy	*Viola tricolor*	Violaceae	Purple, blue, yellow	20-30	Pot culture
44	Aster	*Callistephus chinensis*	Compositae	Pink, rose, purple	30-45	Pot and bed culture
45	Sweet sultan	*Centarrea moschata*	Compositae	Purple, pink, blue	30-45	Bed culture
46	African marigold	*Tagetes erecta*	Compositae	Yellow, orange	45-100	Bed culture
47	French marigold	*Tagetes patula*	Compositae	Red, brown, yellow, orange	20-30	Pot and bed culture
48	Verbena	*Verbena hybrids*	Varbinaceae	Pink, red, purple, blue	30-45	Bed culture
49	Lace flower	*Trachymene caerulea*	Umbelliferae	White. Pink, blue	45-60	Umbrella like clusters
50	Linum	*Linum grandiflorum*	Linaceae	Red, purple	30-45	Bed culture

B. Summer season annuals (Seeds are sown in Feb-March and transplanting in March-April)

S.N.	Common name	Botanical name	Family	Flower colour	Plant height (cm)	Remark
1.	Portulaca	*Portulaca grandiflora*	Portulaceae	Pink, red, yellow, blue, orange	10-15	Pot and bed
2.	Blanket flower	*Gaillardia punlchella*	Compositae	Yellow, brown, orange, scarlet	45-60	Easy to grow
3.	Kochia	*Kochia scoparia*	Chenopodiaceae		60-75	Green leaves gives
4.	Petunia	*Petunia Hybrid*	Solanaceae	Pink, blue, purple, white	20-30	Give good effect
5.	Zinnia	*Zinnia elegans*	Compositae	Violet, orange, white	70-80	Very hardy, easly grown
6.	Sada bahar	*Vinca rosea*		Pink	30-60	Very hardy plant
7.	Coreopsis	*Coreopsis tinctoria*	Compositae	Yellow, scarlet	45-60	
8.	Sunflower	*Helianthus anuus*	Compositae	Yellow, orange	60-120	Can be grown throughout the year
9.	Cosmos	*Cosmos bipinnatus*	Compositae	Pink, orange, yellow	60-70	Bedding purpose

C. Rainy season annuals (Seeds are sown in June and transplanting in July)

S.N.	Common name	Botanical name	Family	Flower colour	Plant height (cm)	Remark
1.	Balsam	*Impatiens balsamina*	Balsaminaceae	Pink, red, rose	60-70	Very delicate
2.	Cocks comb	*Celosia argentea var. cristata*	Amaranthaceae	Red, yellow orange	30-60	Hardy plant
3.	Zinnia	*Zinnia elegans*	Compositae	Violet, orange	70-80	Hardy plant
4.	Amaranthus	*Amaranthus caudatus*	Amaranthaceae	Pink,white	45-60	Grown in pots
5.	Button flower	*Gomphrena* spp.				

44

Edge Plant

Edge plant

Very low growing plants grown only for the purpose of decoration and demarcation and not for screening are known as edging. Live and inert materials used for living of borders flower beds, paths, lawn, shrubbery and herbaceous border are known as edging. Formal edgings are: Bricks, stones, tiles, concrete etc. Informal edging are: Grass verges and crazy paths made up of tiles stones etc.

Foliage plants used for edging

S.N.	Common name	Botanical name	Family	Plant height(cm)
1.	Alternanthera	*Alternanthera amabilis*	Amaranthaceae	3-5
2.	Eupatorium	*Eupatorium cannabium*	Compositae	4-6
3.	Iresine	*Iresine herbstii*	Amaranthaceae	3-5
4.	Duranta	*Duranta spp.*	Verbenaceace	4-6
5.	Pilea	*Pilea muscosa*	Urticaceae	2-5

45

Lawn

Lawn

Lawn is a natural green carpet, and it is important feature of a landscape and a garden without a lawn is not considered complete. Lawn is an area of land closed mowed grasses it is primarily developed for aesthetic and recreational purpose. A lawn is an integral part of a garden and landscape besides aesthetic and recreational purpose and several other purposes. Lawn provides a place for taking rest after tiring of the day, holding parties, social functions, passive and active recreation. Distinguishing characteristic of turf grasses is the ability to with stand close mowing and still providing a functional dense healthy ground cover. Lawn is also called as heart of garden. J.B. Olcoth designed first experiment in lawn grass at Monchester in USA in the 19th century. Festuca and Agrostis species were used to make turf (Lawn) during 1880 in U.K. In the early 19th century lawn become wild spread out side of parks golf course. Now a day in many metro cities of India lawn has become a cultural of the passion of many homes.

SELECTION OF GRASS

There are mainly two types of grasses are used for planting lawn.

1. **Bermuda grass** (*Cynodon dactylon*)- It is commonly called as doob or hariyali. This is very commonly used for planting lawn due to its fast, growth, hardiness, less water requirement and response to frequent mowing. This makes an excellent turf. It is highly suitable for large area and play grounds on account of its tolerant to water and has good recuperation habit.

   Variety of bermuda grasses- Kalkatiya, Hariyali, Selection 1, 5, 8, Palna, Panam etc.

2. **Korean grass** (*Zoysia japonica*)- It is native of Japan, Korean and Philippine island. This kind of grass is recently introduced in India and has become very popular in short span of time due to its normal growth

and more cold tolerance. It makes cushion like turf. This is highly suitable for smaller areas and home lawns.

3. **Others grasses-** Carpet grass (*Axonopus species)*, Bahia grass (*Paspalum nonatum)* and Buffalo grass (*Buchloe dactyloides)* etc. are planting for lawn making.

PURPOSE OF A LAWN

- It is an important element in garden.
- It leads to unity in garden design.
- It is natural green carpet and is the carpet floor of outdoor room.
- It is a heart of garden and the center for social life.
- It gives restful appearance to the eyes through its green outlook all the time.
- Prevent soil erosion.
- Enhance ground water recharge.

CHARACTERISTICS OF LAWN GRASS

- Look fresh and green throughout the year.
- Cold and drought resistance.
- It should not be patchy.
- Free from attack of disease and insect.
- Quick growing.
- Should not give fowl or bed order.

PREPARATION OF SOIL

1. Dig soil up to 45cm depth and expose sun May and June.
2. Turn soil 2-3 times remove stone rocks and break big clouds
3. Spread 10-15 cm thick layer of well rotten weed free FYM and thoroughly mixed in soil.
4. Sandy loam soil is best for lawn grasses growing which are pH ranges between 5-6.
5. Irrigation the field thoroughly and allow all weeds to germinate.

6. Remove the entire weed along with roots manually or spread non selective type of herbicide, like Paraquat or Gramaxone at the rate 1-1.5 liter in about 800-1000 liter.

SITE FOR PLANTING LAWN

South, South-East, South-West open and sunny place for most part of the day with adequate water availability.

PLANTING METHODS OF LAWN

Following are different methods of planting lawns:

1. SEEDING METHOD

- This method is common to grow cool season lawn grasses.
- About 25-30 kg seed/ha is mixed in 200-500 kg sand and saw dust and broadcast evenly in prepared field.
- Do light rolling.
- Sprinkle water regularly until seedling emergence.
- Less labour is required but lawn is not even.

2. DIBBLING METHOD

- A small bunch of grass alone with roots and little stem is taken.
- Planting is done a spacing of 10cm apart both row to row and plant to plant.
- Do regular watering until establishment.
- It is done in June to September.
- Lawn develop by this method is quick uniform and with more labour and cost.

3. SPRIGGING METHOD

- Grass rot along with little stem are chopper into small pieces.
- Spread this over prepared field during rainy season.
- Do small raking to mix grass in soil.
- Do light rolling.
- Do laboural watering with sprayer.
- Do moving after 70-80 days.

4. TURFING METHOD

- Small pieces of well prepared lawn or turf are cut into square or rectangular shape. Preferable plants are planted in polythene seat.
- Fix these in thoroughly prepared field.
- Do heavy rolling.
- Lawn prepared in clean and weeds free.
- Quickest method of lawn raising.

5. PLUGGING METHOD

- The planting of 5-10 cm diameter square, circular or plucked shape piece of sod at regular interval is called plugging.
- Care should be taken that plugs or well watered but must not be soggy (muddy).
- There is 10 times more planting than springing.
- Quick establishment of lawn, it is advised to plant plugs closely.

MANAGEMENT OF LAWN

Mowing and rolling- The newly planted lawn is not mowed till has established well. In the beginning the grass should be trimmed with the help of grass cutter (Mower). It is the cutting of lawn grass for maintaining it's the activeness and for maximum utility. Mower discover by "Edwing Budding" in 1830 in England. Light rolling should be done on dry ground to suppress the upright growth to anchor the grass firmly in the soil and to keep the level of ground. In rainy and winter season mowing is required at an interval of 7-10 days. During spring season mowing is required at the interval of 15 days whereas, during summer it is done at monthly interval.

Manure and fertilizers- The regular application of fertilizers keeps the grass to grow luxuriantly and maintains the lush green colour of the lawn. Sun hemp is very good green manure before lawn planting. In 30m^2 lawn 3-5q. FYM, 10-20kg lime and 10-20 kg SSP. Broadcast a mixture of 50-60gm/m^2 or 1.5kg/30m^2 (2 CN: 1 SSP: 1 K_2SO_4) twice in Feb-March and Aug-Sept. and spray of urea at 0.3%

Irrigation- Water requirement of lawn depend upon season, type of soil, grass used, weather and planet. Grasses are surface feeder and hence adequate watering should be done frequently. It should be done before wilting is internal water stress. Increased watering inters well result in deeper root development, then by decreasing water requirement. Irrigation required up to 5-15cm on depth at 8-10 days interval is ideal as frequent light watering is harmful. Korean grass needs more frequent watering than Calcutia grass.

Weeding- Regular mowing checks weed growth by removing upper portion of the weeds and starvination of root. But still there may be large population of weeds which grows fast and needs removal. For controlling broad leaved weed spread 2-4-D at 0.005% and for narrow leaved weed spray Atrazene @ 1.5 kg/ha in 1000 litter. In lawn of Korean grass spray Benefine or Sylvex at 0.1%.

Scraping of lawn- It is done to renovate the old lawn when it becomes old and grass become compact. After 3-4 years during summer month i.e. during June, lawn should be scraped completely with the help of khurpa and raking should be done both ways. Before the starts of rains top dressing of mixture consisting of garden soil, sand and sieved leaf mould (1:2:1) should be applied to cover the upper 3-5cm.

Disease

- Damping off, root rot, gray leaf, mould, leaf spot, powdery mildew, fairy ring are most important disease of lawn.
- Bordeaux mixture 4: 4: 50 at fortnightly interval.
- Dithane-M-45 (0.1 %) + Bavistin (0.1%) at fortnightly interval.

Insect

- Termite (0.03% Chloropyrophase), cut worm (Spinopad at 0.05%) and root grub (Phorate 10-12 kg).

Silent point for lawn

- Avoid water standing in rainy season.
- Remove all dead or dry leaves falling during autumn season.
- Do raking in lawn twice, once before rain and 2nd after rains.
- Do thinning as at when require.

46

Garden Design

Purpose-Planting of foliage and flowering trees, shrubs, climbers, annuals, perennials and lawn in a piece of land by adopting certain principles is termed as garden design. The objective of a garden layout by adopting a design to provide recreation, enjoyment and education to the society. Natural flora, religion, topography of land and climatic had played a major role in the development of garden designs. Basically there are two garden designs. In one the plantation of shrubs, trees, climbers etc. and lawns is done by adopting a geometrical design and maintaining bilateral balance whereas, in the other one the plantation is done by taking in account the topography and terrin of land, containing open lawns and cluster of trees. The former is called formal are artificial garden while the later is called as informal or landscape garden.

Main features of gardens

(A) Feature of formal garden

- First plan is made on paper and then land is selected accordingly.
- Land is leveled.
- Symmetrical design.
- Geometrical (Square, Rectangular, Circular beds border).
- Raods and path cut a right angle.
- Balance is symmetrical as a same feature replicated on both side of center axis.
- Hedge, edge, topiary trimmed.
- Tree can be selected as individual feature.
- They are usually small to medium in size.
- It has east and west orientation Mughal garden, Persian, Italian and French American.

(B) Feature of informal garden

- Plan is forced to fit the land.
- Main aim is to capture natural scenery.
- Land is not labeled.
- A symmetrical design.
- Non geometrical beds and border, untrimmed hedge, edge, and topiary.
- Individual plant as not selected as feature.
- Much more variety of elements is use.
- Animal like, umbrella, seats, cassettes, rockery, water pull etc. Such as Japanese, Chinies, English.

(C) Feature of free style garden

- A new approach to gardening that allows developing a garden with what on hand.
- This style combine the god point of both formal and informal as well as naturalistic feature and aesthetic mixed to create a picturesque effect.
- This style is suited to almost all situations.

(D) Feature of wild style garden

- Wild styles gardening no rules are followed but aim is to make the garden is beauty and natural "William Robinsons".
- Aim is to make garden more beautiful and natural.
- Such gardens are laid out for more agreeable communication in nature.
- Wild variety of trees, shrubs and creepers are used in natural lays.
- No formal rules are fallowed and allowed to give in their natural shapes.

Principle of garden design

1. As far as possible the design should be simple.
2. It should provide comfort, recreation and educational value.
3. The entire garden should not be visible at a glance and the visitor has to walk to see other components of the garden.
4. Avoid overcrowding, select hardly trees and shrubs which remain in bloom for a longer period of time and provide colour to the garden.

5. As far as possible avoid long and string paths.
6. Arrangements of shrubs, trees, climbers should planned to provide either contrast or harmony.

ESSENTIAL COMPONENTS OF FORMAL AND INFORMAL GARDENS

(A) Formal garden

1. Lawn, 2 Road and Path, 3. Hedges and edges, 4. Annual beds, 5. Birds path, 6 Sun dial, 7. Rosery, 8. Topiary, 9. Shrub and shrubbery border, 10. Rockery, 11. Arbour, 12 Water fall and fountain.

(B) Informal garden

1. Shrubs and trees, 2 Lilly pool, 3. Stone and statue, 4. Walls and steps, 5. Road and paths, 6. Shrub and Shrubbery border, 7. Large open lawn, 8. Rockery, 9. Lantern

(C) Picturesque design

1. Grass, 2 Bulbous plants, 3. Trees and shrubs, 4. Climbers

Material and Methods- Drawing sheet, pencil, rubber, scale, pastel colour, rope and tape

Procedure-The objective of this experiment is to provide opportunity to design from various type of garden on drawing sheet i.e. bungalow, park, college garden, roadside park on national park. It is pre-requisite that the plan be prepared on a picce of paper. The shape and size of the land on which the garden is to prepare should be assumed for drawing the plan on paper by assuming a scale. Decide the style of garden. Demarkets the various components such as road, path, terrace, pool, rocky etc. Indicate the position of shrubs and trees. The trees and shrubs be coloured with natural colour of the flowers. Number each component of the garden and provide explanations for each in the bottom. For better understanding a few garden designs are given (see plates 19). For list of plants it is suggested that students should consult standard horticulture books (See suggested references and plates).

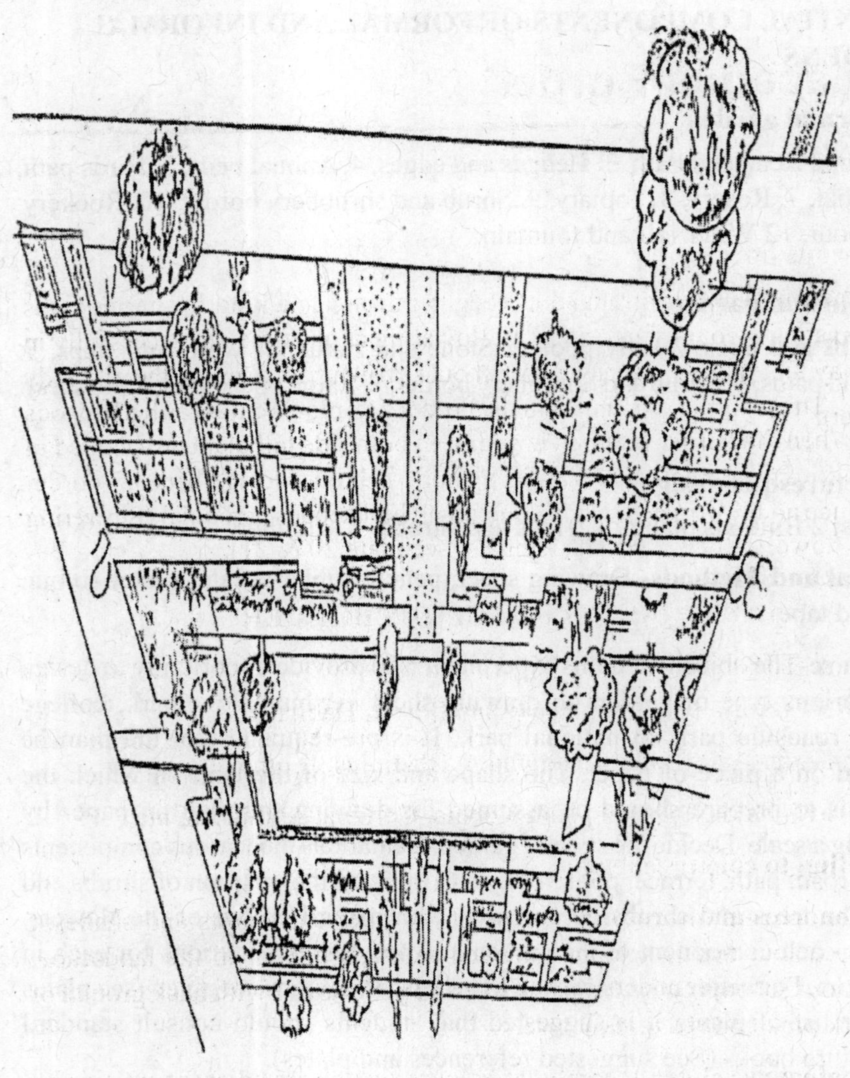

Plate. 19: Part of Mugal Garden at Rashtrapati Bhawan

47

Herbaceous Border

Herbaceous border

In this system plant are arranged in irregular group of a kind for harmonious are contrasting colour effect, either all flowering at one time or successfully in such a way that those flowering latter should grow up and screen the all ready flowered. The planting of annuals in the border of a plot is called as herbaceous border. When the border is to be viewed from both side tall plant are planted in center followed by medium and dwarf planting in both side. Herbaceous border may include all the herbaceous plant including herbaceous perennial (Flowering and non flowering) bulbous and annuals (see plate 20 & 21).

PLANT SUITABLE FOR HERBACEOUS BORDER

A. According to hight

1. **Tall-** Corn flower, Holly hock, Sun flower, Dahlia etc.
2. **Medium-** Aster, Salvia, Geranium, Gladiolus, Lillium etc.
3. **Dwarf-** Phlox, Iris, Ice plant, Kochia, Gerbera, Anthurium etc.

B. Acording to colour scheme

1. **Monochromatic colour-** It consists of different tempts and shade of one colour and is seldom achieved in its pure from in the landscape. Colour scheme could include white and pink flower with back ground of dark pink and red brick house.
2. **Analogous colour-** It combines coloures which are adjacent are side by side on colour wheel an analogous colour scheme includes green , blue green, green blue, blue violet blue.
3. **Complementary colour-** Red and green would be complementary colour a complementary colour scheme may be achieved by using plants with green foliage against a red brick house. Transition is gradual change transition in colour can be illustrated by redial sequence on the colour wheel (Monochromatic colour scheme).

Nowadays the herbaceous border may need to make use of some structural planting from roses or topiary to earn a place in the garden and look attractive for longer period eg. Opium poppy (*Papaver somniferum*) and fox gloves (*Digitalis purpurea*). Where on the area of landscape desien are covered 20% with herbaceous border.

Planning of flower border- The first step in planning the materials for an all season, mixed perennial, border is to select key plants for line, mass, colour and dependability. Line is the outline of a plant, mass is its shape or denseness and dependability refers to its ability to remain attractive with a minimum of problems. The most attractive flower borders are those which are located in front of a suitable background such as a fence, shrubbery or a building. In some cases tall flower such as hollyhocks or sunflowers may serve a dual purpose as flower in the border and as background plants. Annuals or perennials flowers of medium height may serve as background plant for a short border planting eg. Tulip, zinnia, daffodil, salvia, pansy, celosia, ageratum, impatiens, dwarf marigold and gazania.

Tall plants should be selected for the back parts of the bed, with medium height species in the middle and dwarf varieties along the front as edging.

Plate 20: According to height

Plate 21: According to colour scheme

48

Shrubbery Border

Shrubbery- An area of cultivated shrubs in a park or garden is known as shrubbery or any area in the garden fully solely devoted to the shrub plantation is called shrubbery. These shrubs are maintained in formal way therefore require regular training, pruning and clipping. Maintenance of shrubbery is easy and become less costly if planted properly considering following point (See plate 22).

- Selection of hardy shrubs.
- Shrubs having profuse flowering and beautiful foliage.
- Selection of shrub according to soil.
- Shrub having slow growing habit and required less pruning and management.

PRINCIPLES AND PLANNING OF SHRUBBERY BORDER

- Shrubs should be planted in east, south or south-west direction to achieve best result.
- In a large garden shrubbery should be arranged in more area to reduce space and maintenance cost of garden.
- Shrubbery placed in the front of tall tree or along the conifers looks very appearing.
- Flower and foliage colour should be visible from a distant place when tall, medium and dwarf shrub are arranged in the shrubbery.
- Along the path, drives terrace lawn and in front of house small shrub should be planted. Whereas, in front of big tree taller shrubs should be planted for picturesque effect.

CLASSIFICATION OF SHRUBS

- According to height
- According to colour scheme

- According to beauty of plant
- Acording to requirement of sunlight

(1) According to height- In front of an object i.e. Tree, big building. Taller shrubs are planted first, then medium shrub to be planted after that smaller shrubs are place in the shrubbery.

(a) **Dwarf-** Up to 1 m. eg. *Barleria cristata* (violet blue), *Crossandura undulaefolia* (Yellow), *Erathemum laxiflorum* (purple-rose), *Jasminum sambac* (white), *Lantana camera* (blue) etc.

(b) **Medium-** 1.0 to 2.5m. eg. *Acalypha hispida* (Red), *Allamanda neriifolia* (Yellow), *Cestrum nocturnum*, (White), *Cestrum diurnum* (White), *Jasminum multiflorum* (White), *Lantana camara* (White) etc.

(c) **Tall-** 2.5 to 4m. eg. *Buddelia asiatica* (white), *Clerodendrum inerme* (white), *Gardenia jasminoides* (white), *Hamelia patens* (Red), *Hibiscus rosa chinensis* (Red) etc.

Arrangement of shrubbery- Shrubbery is arrange basically in two way ie.

(a) **Single face shrubbery border-** In this case plantation of shrub is done in front of trees on boundary. Shrubs are arranged according to height then the taller shrub in the last then medium and after that smaller shrubs should be planted.

(b) **Double face shrubbery border-** Its observed from both sides in this case taller shrubs are planted between the two medium shrubs besides with smaller shrub (up-to 1m ht.) are planted in both side.

(2) According to colour scheme- These shrubs have attractive colour light yellow, white, golden yellow, pink, scarlet, crimson, rose red, violet, blue etc. there are three common colour schemes are given below.

(a) **Monochromatic colour scheme-** Massing of single or one colour is called monochromatic colour scheme eg. White- *Jasminum sambac* (dwarf), *Cestrum nocturnum* (medium), *Gardenia joaminoids* (tall) and Yellow- *Galphimia gracilis* (dwarf), *Allamanda herifolia* (medium), *Thevetia peruviana* (tall).

(b) **Analogous colour scheme-** It is also called harmonious colour scheme in this scheme shrubs are planted with closely related colour like white, pink, red colour flower or vice versa.

(c) **Contrast colour scheme-** It is also called complementary colour scheme. Two opposite colour shrubs are planted in this scheme like blue-orange, red-green and yellow-violet.

(3) According to beauty of plants- These shrubs have attractive in flowering, foliage, variegated, fruit and fragrance quality.

(a) **For flowers-** *Barleria cristata*, *Bougainvillea species*, *Crossandra undulaefolia*, *Hibiscus rosa-sinensis*, *Jasminum pubescence* etc.

(b) **For foliage-** *Codiaem varigatum, Euphorbia cotinifolia, Manihot species,* Aralia etc.

(c) **For variegated foliage-** *Duranta plumier, D. varigata. Manihot utilissima, Tebernaemontana coronaria* etc.

(d) **For flower and foliage-** *Acalypha hispida*, Bougainvillea (Archana, Bhabha), Hibiscus (Snow queen and Redhot), *Hamelia patens* etc.

(e) **For fruits-** *Carissa carendus* (Dark red), *Citrus Japonica* (Yellow), *Duranta macrophylla* (Yellow), *Rauwolfia canscess* (Red) etc.

(f) **For fragrance-** *Cestrum nocturnum* (Rat-Ki-Rani), *Cestrum diurnum* (Din-Ka-Raja), *Jasminum auriculatum*, *Jasminum grandiflorum*, *Jasminum sambac* etc.

(4) According to requirement of sunlight

(a) **For sunny situation-** Bougainvillea, *Hibiscus rosa-sinensis*, *Jasminum sambac*, *Jasminum grandiflorum, Lanata camara, Murraya exotica* etc.

(b) **For partial or semi shade-** *Jatropha rosia, Mussaend philipica, Maglolia grandiflora, Nandina domestica* etc.

(c) **For both partial and sunny shade-** Acalypha, *Cestrum nocturnum, Chrosandra sps., Hemelia patens, Thunvergia errecta* etc.

Plate 22: Shrubbery border according to heights

(3) **According to beauty of plants**: [illegible] are also [illegible] and fragrance again.

(a) **For flowers**: [illegible]

(b) **For foliage**: [illegible] Aralia etc.

(c) **For variegated foliage**: [illegible] etc.

(d) **For flower and foliage**: [illegible] Bhabhal, Hibiscus [illegible] etc.

(e) **For fruits**: [illegible] Rose [illegible] etc.

(f) **For fragrance**: [illegible] Raat Ki Rani [illegible] etc.

(4) **According to requirement of sunlight**

(a) **For sunny situations**: [illegible] etc.

(b) **For partial or semi shade**: [illegible]

(c) **For both partial and heavy shade**: [illegible] etc.

Plate 12: [illegible]

Section D
Spices, Medicinal, Aromatic and Plantation Crops

49*

Spices and Condiments

*Table starts from next page.

Description of important spices and condiments

S.N.	Common name	Botanical name	Family	Origin	2n No.	Type of inflorescence	Type of fruit	Propagation	Plant part used	Chemical content
1.	Black pepper	*Piper nigrum* L.	Piperaceae	Indo-Burma	52	Pendent spike	Drupe	Shoot cutting	Fruit, seed	Piperine
2.	Small Cardamom	*Elettaria cardamomum*	Zingibraceae	India	48	Panicle	Capsule	Suckers	Fruit, seed	Cineol
3.	Ginger	*Ginger officinalis*	Zingibraceae	South- east Asia	22	Spike	Capsule	Rhizomes	Rhizomes	Zingiberene
4.	Turmeric	*Curcuma longa*	Zingibraceae	South- east Asia	62	Spike	Capsule	Rhizomes	Rhizomes	Curcumine
5.	Cinnamon	*Cinnamomum verum*	Lauraceae	Srilanka	24	Paniculate cymose	One seeded berry	Rooted cutting, air layering, seedling	Bark	Cinamaldehyde
6.	Nutmeg	*Myristica fragrans*	Myrsticaceae	Indonesia	42	Umbellate, cymes	One seeded drupe	Epicotyl grafting	Seed, aril	Terbein
7.	Clove	*Syzygium aromaticum*	Myrtaceae	Indonesia	22	Terminal cymes	Single Seeded drupe	Seed	Unopened flower bud	Eugenol
8.	All spice	*Pimenta dioca*	Myrtaceae	West indies	22	Racemose cymes	Berry	Seed, air layering	Fruit, seed	Phenol
9.	Capsicum	*Capsicum annum*	Solanaceae	South America	24	Solitary axillary	Berry	Seeds	Fruit	Solanine
10.	Curry leaf	*Murraya koenigi*	Rutaceae	India, Srilanka, and other south india country	18	Terminal cymes		Seeds	Leaf	Koenigine
11.	Asafoetida	*Ferulla foetida*	Apiaceae						Oleoresin gum	Ferumine

Contd.

12.	Saffron	*Crocus sativas* L.	Iridaceae	india	24			Corms	Stigma	Cicrocrocin
13.	Sweet flag	*Acorus calamus*	Araceae					Rhizome	Rhizome	Acorin
14.	Dill	*Anethum graveolens*	Apiaceae	South Eurasia and western asia	22	Umbel	Schizocarp	Seeds	Fruit	Carvone
15.	Coriander	*Coriandrum sativam*	Apiaceae	Mediteranean region	22	Compound umbel	Schizocarp	Seeds	Leaves, seed	Linalool
16.	Cumin	*Cuminum cyminum*	Apiaceae	Mediteranean region	14	Compound umbel	Schizocarp	Seeds	Seeds	Cuminol
17.	Fennel	*Foeniculum vulgare*	Apiaceae	Mediteranean region	22	Umbel		Seeds	Fruit	Limone
18.	Aniseed	*Pimpinella anisum*	Apiaceae	Mediteranean region					Fruit	Anithol
19.	Bishops weed (Ajwain)	*Trachyspermum amni*	Apiaceae	Mediteranean region	18	Compound umbel		Seeds	Fruit	Thynol
20.	Vanilla	*Vanilla planifolia*	Orchidaceae	America	32	Raceme	Capsule	Stem cutting	Fruit, pod	Vanillin
21.	Fenugreek	*Trigonella foenumgraceum*	Fabaceae	Europe	16	Racemose	Pod	Seed	Seed	Diosgenin
22.	Kokum	*Garcinia indica*	Guttiferae	India	24				Fruit	β hydroxy acetic acid
23.	Celery	*Apium graveolens*	Apiaceae	Southern Europe	22	Umbel	Schizocarp	Seed	Leaf and fruit	Carvone
24.	Garlic	*Allium sativam*	Alliaceae	Central asia	16	Umbel		Cloves	Bulbs	Allicin (diallyldisul phide oxide)

50*

Medicinal Plants

*Table starts from next page.

Description of important medicinal plants

S.N.	Common name	Botanical name	Family	Origin	Plant part used	Chemical content	Uses
1.	Belladonna	*Atropa belladona*	Solanaceae	Europe	Leaves	Atropine Hyoscyamine	Neurologic pain, ampholinergic property Hyoscyamine used as truth confessor in criminological investigations
2.	Foxglove	*Digitalis purpurea*	Scrophulariaceae	Europe	Leaves	Digitoxin	Cardiotonic property. Used in cure heart diseases
3.	Henbane	*Hyoscyamus niger*	Solanaceae	Europe	Leaves	Hyoscyamine	Used to cure Asthama and whooping cough. Antispasmodic and Anticholingenic property
4.	Senna	*Cassia angustifolia*	Leguminoceae	Europe	Leaves	Sennosides A,B,C,D	Laxative property, constipetion
5.	Dill or sowa	*Anethum graveolens*	Apiaceae	Europe	Seeds	Carvone	Oil is used to make gripe water
6.	Ashwagandha (winter cherry and Indian Ginseng)	*Withania somnifera*	Solanaceae	Africa	Roots	Withanine somniferine	Used to make general tonic
7.	Sarpagandha	*Rauvolfia serpentina*	Apocynaceae	India	Roots	Serpentine Reserpine Saponins	Used as treatment for hypertension as a sedative (reduce nervous exitement)
8.	Safed musali	*Chlorophytum boribilianum*	Liliaceae	Africa	Roots	Saponins	To make vital tonics, to cure general diabeties. IInd silajeet
9.	Liquorice or Mulathi	*Glycyrhiza glabra*	Leguminoceae	Iraq	Roots	Glycyrrhizin	Used to cure intestinal and peptic ulcers 150 times sweeter than sugar
10.	Periwinkle or Vinca	*Catharanthus roseus*	Apocynaceae	West indies	Roots and leaves	Amacline, vincristine, vinblastine	Used as tranquilizer (blood pressure control), cancer theraphy

Contd.

11.	Opium (poppy)	*Papaver somniferum*	Papavaraceae	Europe	Fruits (capsule)	Morphine, codine, narcotine	God of sleep, painkiller, analgesies (reduces pain) and hypotonic effect, used to cure leukemia and found linaleic acid
12.	Isabgol (psyllium)	*Pantago ovata* (stemless)	Plantagonaceae	West indies	Husk, seed	Mucilage	Laxative, soothing and cooling agent, used against irritation in gastrointestinal tract
13.	Medicinal solanum	*Solanum khasianum*	Solanaceae	Newzealand	Fruits	Solasodine	Major source of steroid in India
14.	Aloe	*Aloe vera, A. barbadensis*	Liliaceae		Leaves	Aloin oil	Skin tonic and herbal cosmetics
15.	Medicinal yam	*Dioscorea floribunda*	Dioscoreaceae		Roots	Diosgenin	Production of sex hormones and contraceptic pills
16.	Neem	*Azadirachta indica*	Meliaceae	India	Fruits & leaves	Azadiractin	Diabeties treatment
17.	Cinchona	*Cinchona spp.*	Rubiaceae	South America	Bark	Quinine	Treatment to malaria
18.	Datura	*Datura innoxia*	Solanaceae	Mexico	Whole herb	Hyoscine tropane	Preanasthetic in surgery, in relief of withdrawal symtomps in morphine
19.	Ipecac	*Cephaelis ipecacuna*	Rubiaceae	Brazil	Roots	Cephaline, emetin	To reduce vomiting, use against amoebeiosis
20.	Long pepper	*Piper longum*	Piperaceae	India	Roots	Piperine	Improve appetiti, laxative
21.	Guggal	*Commiphora wightii*	Burseraceae	Africa	Oleo gum resin	Guggulipids	Treatment of obesity, anthritis

51*

Aromatic Plants

*Table starts from next page.

Description of important aromatic plants

S.N.	Common name	Botanical name	Family	Origin	Plant part used	Oil content	Uses
1.	Davana	*Artimesia pallens*	Asteraceae	India	Leaves	Hydrocarbons	Delicate fragrance for floral decoration. Bouquets, cosmetics
2.	Rose geranium	*Pelargonium graveolens*	Geraniaceae	South Africa	Leaves and flower	Geraniol rhodinol	Perfumes, powder, creams, body lotions
3.	Japanese mint	*Mentha arvensis*	Labiatae		Leaves	Menthol, carvone, linalool	Scenting in the supari
4.	Bergamot mint	*Mentha citrate*	Labiatae		Leaves	Carvone	
5.	Oil bearing rose	*Rosa damascena*	Rosaceae		Flowers	Citronellal, geraniol	Auto of rose, ruha gulab, rose oil
6.	Lemon grass (kochin oil)	*Cymbopogon flexuosus*	Gramineae	India	Leaves	Citral, farnesol	Starting material for ionones and Vitamin A manufacturing
7.	Patchouli	*Pogostimon patchouli*	Laminaceae		Leaves	Patchouliol	Fixative property
8.	Vitever grass	*Vetevaria zizanoides*	Viteveraceae	India	Roots	Viteverol	Carminative property
9.	Java citronella	*Cymbopogon winterianus*	Gramineae	Ceylon	Leaves	Citronellal	Mosquito repellents, deodorants, scented soaps
10.	Palmrosa grass	*Cymbopogon martini*	Gramineae		Leaves	Geraniol	Soaps, perfumery
11.	Kewda (serimpine)	*Pandanas fassicularis*	Pandanaceae	Tropical	Flowers	Lupilin	Keweda water (rooh or attar)
12.	Hops	*Humulus lupulus*	Cannabinaceae	Africa	Flowers		
13.	Ambrette or muskdana	*Abelomoschus moschatus*	Malvaceae	India	Seeds	Fernesol	Musk odour used in incense stickes, panmasala, perfumery, cosmetics, scent

Contd.

14.	Jamalagota	*Croton tiglium*	Euphorbiaceae		Fruits		Violent purgative
15.	Celery	*Apium graveolens*	Umbellifereae	Italy	Seeds	Limonine	Seeds: spice seed oil seasoning, flavouring sauces, purees
16.	Indian basil	*Occimum basilium*	Lamiaceae	Africa	Leaves and inflorescence	Methyl chavicol	Flavouring foods
17.	Rosemary	*Rosemarius officinalis*	Lamiaceae	Europe	Whole herb	Camphene, cineol	Anticancer and antioxidant property
18.	Melissa	*Melissa officinalis*	Lamiaceae		Whole herb	Citrol, nerol	Perfumery, cosmetics

52*

Plantation Crops

*Table starts from next page.

Description of important plantation crops

S.N.	Common name	Botanical name	Family	2n	Origin	Plant part used	Remarks
1.	Tea	*Camellia sinensis*	Theaceae	30	China	Leaf	Queen of beverage crop
2.	Coffee	*Coffea* spp.	Rubiaceae	22	Ethiopia/ Central America	Berries	King of beverage crop
3.	Coconut	*Cocos nucifera*	Araceae/palmae	32	South East Asia	Nut	King of species/ kalpavriksha
4.	Arecanut	*Areca catechu*	Araceae	32	Indonesia	Nut	Supari/betel nut
5.	Cocoa	*Theobroma cocoa*	Sterculiaceae	20	Amazon valley of South America	Pods	Food of god
6.	Cashewnut	*Anacardium occidentale*	Anacardiaceae	42	Brazil	Nut	Plough crop/dollar earning crop
7.	Rubber	*Hevea brasiliensis*	Euphorbiaceae	36	Brazil	Latex	Natural rubber/para rubber
8.	Palmyrah palm	*Borassus flabellifer*	Palmae	36	Tropical Africa	Fruit	Kalpaka virucham
9.	Oil palm	*Elaeis guineensis*	Araceae	32	West Africa	Fruit	Crop for future
10.	Betel vine	*Piper betle*	Piperaceae	26	East Malasia	Leaves	Pan

Section E
Post Harvest Technology

53

Solution for Analytic Procedures

Standard solution: Standard solution is defined as a solution whose strength is known. e.g. 1 N solution.

Molar solution: Molar solution can be defined as a solution containing the number of moles of a solute (substance) i.e. molecular weight in gram per litre of the solution. e g. Mol. wt. of NaCl 23+35.5 58.5,

So 58.5 g NaCl per 941.5 ml of water= 1 M NaCI solution. Molarity is generally designated by M.

Molal solution: Number of moles of a substance/solute per litre of solvent makes a molal solution. Mobility is generally designated by m.

Normal solution: A normal solution is defined as one, which contains a solute/ substance equal to its equivalent weight in grams per litre of the solution. N designates normality

Percent solution: Number of parts of a solute per 100 ml of a solution represents a per cent solution.

Parts per million (ppm) solution: Number of milligrams of a solute per litre of the solution represents a ppm solution.

Buffer solution: Buffer solution is characterised by a definite pH value and which does not allow a sudden change in pH oven with the addition of a strong acid or a base.

Indicators

1. **Phenolphthalein :** pH range 8.3 to 10.0 per 100ml Dissolve 1 g in 100 ml of neutral ethyl alcohol and water. Use 1 drop per 100 ml solution.
2. **Methyl red:** pH range 4.4 to 6.0 Dissolve 1 g per 100 ml of 95 per cent ethyl alcohol. Take reading immediately after adding
3. **Methyl orange:** pH 2.9 to 4.0 Dissolve 0.5 g in 1000 ml water.
4. **Bromophenol blue:** pH 3.0 to 4.6 Dissolve 0.1 g in 100 ml water.

5. **Starch solution:** Dissolve 0.5 g in 15 ml. Add 100 ml hot water to it. Boil for 1-2 minutes.

6. **Methyl blue:** pH 3.0 to 4.2. Dissolve 0.5 g in 1 litre water.

Cleaning solution: Dissolve 40g sodium or potassium dichromate to 150 ml of water and then add 250ml sulphuric acid.

Precaution: This solution should always be prepared over a sink.

54

Determination of TSS

Purpose: The estimation of total soluble constituents gives the approximate amount of water soluble substance present in the sample. Among the various soluble substances the amount of sugar is 80-85 percent. The TSS value is roughly considered to be the amount of sugar and soluble minerals present in fruits and vegetables. The total soluble solid content can be determined by hand refrectometer.

Material and equipments: Hand refrectometer, juice, muslin cloth or tissue paper and glass rod.

Description and use of the apparatus: With the help of this instrument, the percentage TSS of plant juices can be determined quickly. It work on the principle of total refraction of plant product (juice)

Note: Total soluble solids are measured at room temperature with the help of a Hand (0-32)/Abbe (0-100) refractometer equipped with a percent scale. Temperature correction if necessary should be made with 20^0C as standard temperature. The refractometer should be cleaned after each observation.TSS is expressed in per cent or ^{0}Brix. (photograph of hand refrectometer are given in tools and equipments chapter)

55

Determination of Total Solids

Principle: A weighed sample is dried in an oven at 60-70^{0}C up to a period of time when weighing made at 2 hour intervals do not differ by more than 2-3 mg.

Procedure: Take 20-25 g of sample in a large dish. Put it in vacuum oven at 60-70^{0}C temperature and 26 inches vacuum till the subsequent weighing of the sample at 2 hour intervals do not differ by more 2-3 mg general overnight drying is sufficient for most of fruit samples.

56

Determination of Ash

Purpose: The ash content of plants represents the deposition of minerals elements in the plant or in its parts. The value of ash content in the plant tissue indicates the health of the plant. Plant ash is also used both in qualitative and quantitative estimation of mineral elements.

Material and Equipments: Crucible, oven, chemical balance and weight box.

Principle: Died sample is ignited at a sufficient high temperature (more than 500°C) to a white ash

Procedure- Select fresh leaves which are fully developed mature, free from diseases and pests. Leaves fully exposed to sunlight should be preferred. The leaves of cabbage or tomato or mustard be tried. Wash the leaves under running water. Chop the leaves into smaller pieces with the help of a stainless steel knife. Weigh 100 gm of chopped leaves on a chemical balance. Now transfer the weighed material in 25 ml beaker. Put the beaker containing the chopped leaves on oven. Maintain the temperature of oven between 102-103°C. After four to six hours remove the beaker from the oven and cool the contents in a dessicator. On cooling weigh the contents. The contents of the beaker should be heated, cooled and weighed two to three times till a constant weight is obtained. The difference between the fresh and dry weight represents the water content of the sample.

Now transfer the contents of the beaker in a crucible of known weight. Put the crucible containing the dry matter in a Muffle furnace. Maintain the temperature of furnace at 400°C. Continue the heating till a white or grey coloure ash is formed. Remove the crucible from the furnace and cool it in a dessicator. Weigh the ash along with crucible. Substract the weight of crucible will give the weight of ash.

57

Water in Soluble Solids

Principle: To determine water in soluble solids, a known quantity of sample is boiled in water and the insoluble material is collected on a weighed filter paper. The insoluble mass is then washed with hot water and dried in oven.

Procedure: Take two sets of weighed quantity of sample in a 500 ml beaker and add about 200 ml. hot water. Boil it for about 20 minutes and then transfer one set to 250 ml volumetric flask and make the volume after cooling at 20°C. Filter the sample from beaker and volumetric flask separately through whatman No.4 filters (washed and dried in oven for 2 hours at 100°C, cooled and weighed in covered weighing dish). Wash the insoluble solids with hot water. Now put the filter paper into original weighing dish and dry in the oven at 100°C for about 12 hours and cool in desiccator. Weigh it again.

Water in soluble solids (%) = Dry wt. of insoluble material x 4

58

Determination of Acidity

Purpose: The organic acids present in fruits and vegetables influence the flavour, colour, keeping quality and maturity. The maturity standards are determined by the acid and sugar ratio. The total acidity (citric acid) is determined by titrating the juice against standard alkali solution.

Equipments and chemicals: 10 ml., pipette, 25ml. Burette, small glass funnel, dropper, juice extractor, distilled water, 100ml. volumetric flask, burette stand, beaker, N/10 sodium hydroxide, filter paper, phenolphthalein indicator (one gram of phenolphthalein in 100ml. of 95% ethyl alcohol)

Principle: Total acidity is determined by titrating the sample extracted in water against 0.1 N sodium hydroxide.

Procedure: Take 5-10 g sample. To this add a little amount of water and mix thoroughly. Now titrate the sample solution against 0.1 N NaOH using phenolphthalein as the indicator. Appearance of light pink colour denotes the end point. The acidity is calculated by using following formula and expressed in per cent.

$$\text{Total acid (\%)} = \frac{1 \times \text{Eq. wt. of acid} \times \text{Normality of NaoH} \times \text{titer} \times 100}{10 \times \text{wt. of sample}}$$

Total acid is expressed in terms of predominant acid in a fruit. For fruits like Citrus, Strawberry, Raspberry, Gooseberry etc., it is citric acid; in apple, peach, plum, apricot, cherries etc. it is malic acid and in grape it is tartaric acid.

59

Determination of Sugar

Principle: The copper in the Fehling solution is reduced to red, insoluble cuprous oxide in the presence of invert sugars like glucose and fructose. The quantity of sugars solution liquids for complete neutralization of a known quantity of Fehling mixture is determined by titration using methylene blue as the indicator.

(Lane and Eynon Method)

Reagents

- **Fehling mixture:** Fehling no.1 and 2 (5 ml each) should be mixed immediately before use for each sample
- **Methyl blue indicator:**
- **Neutral lead acetate (45%) solution**: Dissolve 225 g neutral lead acetate in water and make 500 ml.
- **Potassium oxalate (22%) solution:** Dissolve 110 g potassium oxalate in water and make volume 500 ml.
- 50% HCl
- Phenolphthalein indicator
- NaOH (40%) solution.

Preparation of standard sugar solution: Take 9.5 g sucrose in a 250 ml beaker, and 100 ml water and 10 ml (50%) HCI, Transfer this solution to 100 ml flask and make the volume at 20°C

Neutralization of sugar solution: Take 50 ml invert sugar solution in 250 ml volumetric flask and a about 100 ml water. Add 1 drop of phenolphthalein indicator, add 40% NaOH drop by drop till the pink colour appears. Now add carefully, the 50% HCI drop by drop till pink colour disappears. Make the volume. Titrate against 10 ml Fehling mixture as mentioned under "standard method of titration".

Method of titration: Initially incremental/trial method of titration should be used and after the establishment of correct dilution, standard method of titration should be used.

Incremental method: Take 10 ml Fehling mixture in 250 ml flask. Add 15 ml sugar solution. Keep boiling on a hot plate. If colour remains blue further add 2-3 ml solution slowly till the faintest blue colour remains. Now add few drops of methylene blue indicator and complete the titration by adding sugar solution till colour turns brick red.

Standard method of titration: Pipette out 10 ml of Fehling (5ml of each Fehling No 1 and 2) into duplicate 250 ml flasks. Fill the burette with unknown sugar solution. Add into the flask almost the whole volume required to reduce the Fehling mixture so that not less than 0.5 ml or more is required to complete the titration. Boil the contents for 1-2 minutes add 2 drops of methylene blue indicator complete the titration to the appearance of brick red colour. Add sugar solution drop by drop to complete the titration to the appearance of brick red colour.

Alternate Method

i) **Reducing sugar:** Take 5 ml/g of juice or pulp or powder (in case of dehydrated product) in a glass tube. Add 5 ml of potassium ferricyanide to it. Boil on hot plate for 15 minutes. Then cool under running tap water. Now titrate it against N/100 sodium thiosulphate using starch as indicator. Disappearance of blue colour marks the end point. Simultaneously, run a blank and record the observation.

Calculation

mg. sugar/5 ml juice or extract= [(ml of $Na_2S_2O_3$ used in blank) -(ml of $Na_2S_2O_3$ used in sample) +0.05] x 0.338

ii) **Total sugars:** Take 10 ml sugar extract and 2 ml of HCI (50%) in a flask and keep it for 10 minutes at 68°C. Make volume after hydrolysis of sugars. Neutralize the solution by adding anhydrous Na_2CO_3 soln. till the effervescence stops. The method described under reducing sugars should be employed for further estimation.

Flow sheet for estimation of sugars

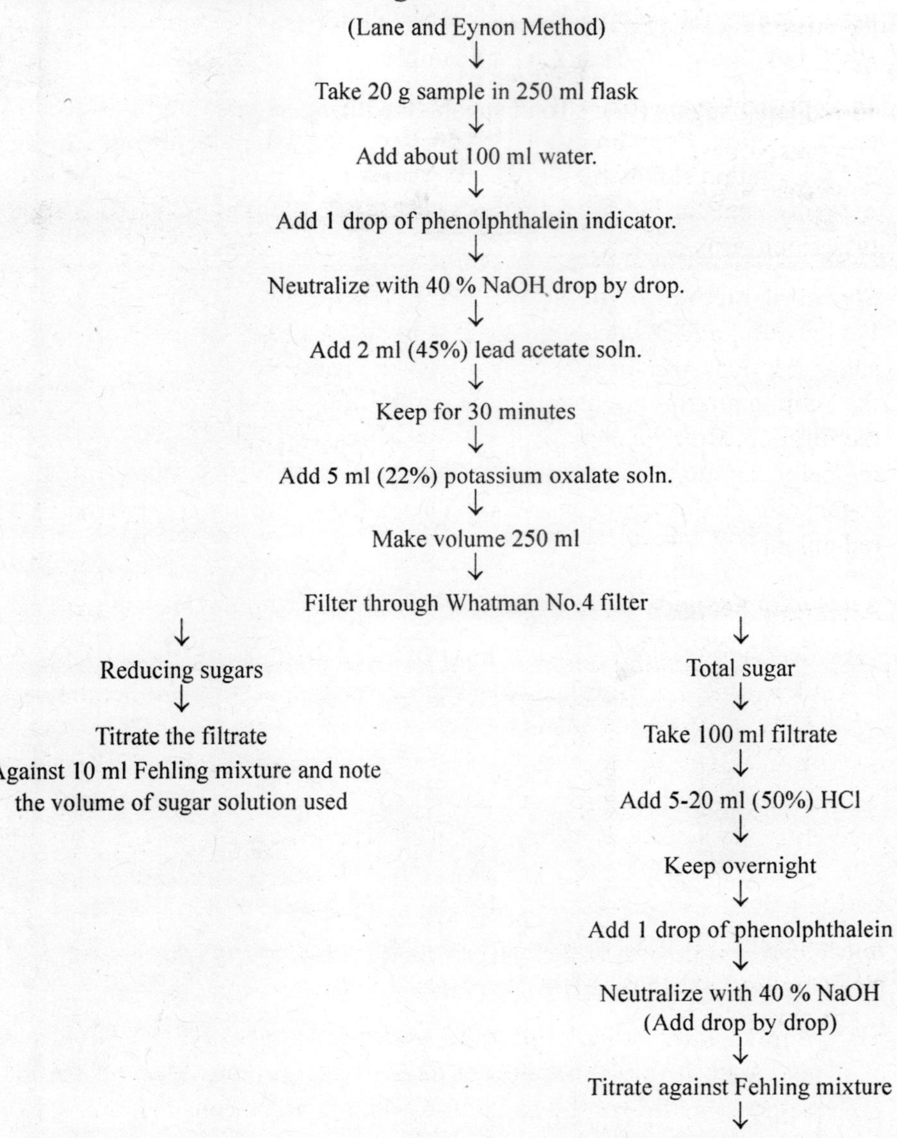

Calculations

$$\text{Reducing sugars (\%)} = \frac{\text{Factor (0.052)} \times \text{dilution} \times 100}{\text{Titer} \times \text{Wt. of sample}} = \text{X (say)}$$

$$\text{Total sugars (\%)} = \frac{\text{Factor x dilution x 100}}{\text{Titer x wt. of sample}} = \text{Y (say)}$$

Non-reducing sugars (%) = Total sugars- Reducing sugars

$$= Y - X = Z$$

60

Determination of Fructose

Fructose, a keto-hexose (called as fruit sugar), is usually accompanied by sucrose in fruits and estimated by the methods of Ashwell (1957).

Materials

Resourcinol reagent: Dissolve 1g resorcinol and 0.25 g thiourea in 10 ml glacial acetic acid. This is stable in the dark.

Dilute HCl: Mix five part of conc. HCl with one part of distilled water.

Standard fructose solution: Dissolve 50mg of fructose in 50 ml water. Dilute 5 ml of this stock soln. to 50 ml for a working standard.

Procedure

1. To 2 ml of the solution containing 20 to 80 mg of fructose, add 1ml of resorcinol reagent.
2. Then add 7 ml of dilute HCI.
3. Pipette out 0.2, 0.4, 0.6, 0.8 and 1 ml of the working standard and make up the volume to 2 ml with water. Add 1 ml of resorcinol reagent and 7 ml of dilute HCl as above.
4. Set a blank along with the working standard.
5. Heat all the tubes in a water bath at 80°C for exactly 10 min.
6. Remove and cool the tubes by immersing in tap water for 5 min.
7. Read the colour at 520 nm within 30 min.
8. Draw the standard graph and calculate the amount of fructose present in the sample using the standard graph.

61

Determination of Ascorbic Acid

Principle: Sample extract in oxalic acid is titrated against standard sodium 2, 6 dichlorophenolindophenol dye to a faint pink colour which persists for 5-10 seconds.

(Indophenols Method)

Reagents

1. **Indophenol dye** (0.04%): Weigh 40 mg sodium 2, 6-dichlorophenolindophenol. Add 150 ml hot distilled water. Then add 42 ml sodium bicarbonate. Cool the contents make 200 ml with water keep in refrigerator.
2. **Metaphosphoric acid** (3%): Dissolve 30g MPA in water and make volume 1000 ml.
3. **Standard ascorbic acid:** Dissolve 100 mg ascorbic acid in 100 ml of oxalic acid. Dilute 10 ml to 100 ml with MPA (1 ml=0.1 mg ascorbic acid).

STANDARDIZATION OF DYE

Take 5 ml standard ascorbic acid and add 5 ml HPO_3. Fill a microburette with dye. Titrate against dye solution to a light pink colour and determine dye equivalent.

Dye equivalent = 0.5/Titer

Preparation of sample: Take 10 g of sample. Add 3% MPA to make volume 100 ml. Mix thoroughly in case of solids or semi-solids and filter. Centrifuge if needed.

Procedure: Take 10 ml of sample filterate and titrate against sodium 2,6-dichlorophenol indophenol dye. Note the titer.

Calculation

$$\text{Ascorbic acid (mg/100 g)} = \frac{\text{Titer x dye equiv. X dilution x 100}}{\text{wt. of sample}}$$

Precautions

1. The dye should not be kept for more than two weeks.
2. Titration should be completed within one minute
3. The dye should be stored in a refrigerator.

62

Determination of Pectin

Principle: Addition of calcium chloride results in the precipitation of pectin as calcium pectate from an acid solution. Calcium pectate is washed with water to make free from chloride and is dried and weighed.

Reagents

1. **Acetic acid:** Normal solution: Add 30 mi glacial acetic acid in 500ml water.
2. **Calcium chloride:** Normal solution: Add 55 g anhydrous $CaCl_2$ in water, dissolve and dilute to 100 ml.
3. **Silver nitrate-**1% solution: Dissolve 5g $AgNO_3$ in water and dilute to 500 ml.

Procedure: Take 25g of the sample in one litre beaker. Add 400 ml water. Boil for one hour. Replace the evaporated water by addition of distilled water. Cool it. Transfer to 500 ml volumetric flask. Filter through Whatman No.4 filter. Take 100 ml of the filterate in two beakers. Add 300 ml distilled water to each. Now add 10 ml 1N NaOH solution and keep over night. Add 50 ml 1N acetic acid. Wait for 5 minutes. Now add $CaCl_2$ solution and keep it for one hour. Boil it for one minute. Now take two Whatman No.4 filters. Wash with distilled water, dry in an oven at 100°C for two hours and then weigh these. Filter the solution through Whatman No. 4 filters. Wash with distilled water to make free from chloride ions. Add a few drops of silver nitrate solution. Put the white precipitates (on filter paper in a petri-dish) in an oven, dry and weigh again.

Calculation

$$\text{Pectin (\%)} = \frac{\text{wt. of Ca pectate x 100}}{\text{wt. of the sample}}$$

Flow Sheet for the Estimation of Pectin

Take 25 g sample in 1000 ml beaker.

↓

Add 400 ml distilled water

↓

Boil for one hour.

↓

Cool

↓

Make volume 500 ml

↓

Add 10 ml 1 N NaOH soln.

↓

Mix well, keep overnight

↓

Add 300 ml distilled water

↓

Add 50 ml acetic acid soln.

↓

Wait for 5 minutes

↓

Add 25 ml 1 N $CaCl_2$ soln.

↓

Wait for one hour

↓

Boil for one minute

↓

Filter through Whatman No.4
(Boiled in distilled water and dried) filter

↓

Dry the filter with pertri-dish in oven

↓

Weigh again

↓

Note Ca pectate

63

Determination of pH

Procedure: Standardize the pH meter with a buffer solution of pH. Take the sample solution in a 100 ml beaker. Put the electrodes in the solution and read on pH meter. After each observation, wipe the electrodes with a piece of cotton soaked in distilled water and rinse the electrodes with water with the help of a wash bottle and dry with filter paper.

pH Meter

It is used for measuring pH since pH is a measure of acidity and alkalinity.

$$pH = \log 10 \frac{1}{H^+}$$

pH ranges from: 0 to 14

0-6.99= acidic

7 = neutral

7.01 14.0 = alkaline

The range of pH is from 0-14 because the product of water (H^+ and OH) is 10 at 25^{0}C. The pH meter is made up of following parts:

1. Rectangle indicator
2. Temperature compensator
3. Selector
4. Measuring electrode
5. Set zero nob
6. Set buffer
7. Glass and reference electrodes.

Procedure: Connect the instrument to power supply with the selector switch "On" in zero position. Wait for 5 minutes. Adjust zero nob to bring the pointer to zero position. Prepare buffer with tablet and adjust pH of buffer with set buffer. Remove the buffer and put the sample and note the reading.

1. Digital pH meter
2. Electrical conductivity meter

64

Determination of Moisture Content

Preparation of the material: Common minute material for analysis in Waring blender for 1 to 2 minutes. Do not over heat. Pass it through No.60 sieve and put is in airtight container.

Procedure: Take 2 to 3 g sample in duplicate in pre-dried and pre-weighed 7 cm diameter aluminium dishes. Remove the dish cover and put them in vacuum oven and dry for 6 hours at 88°C and 26-22 inch vacuum. During this process, admit to oven a slow current of air (2 bubbles per second) dried by passing H_2SO_4 replace the cover, cool it in desiccators and weigh again.

Calculations

$$\% \text{ Moisture} = \frac{\text{Loss in wt.} \times 100}{\text{Weight of sample}}$$

65

Canning

Canning: Cannings the process by which sealing of the food product and sterilizing them by heat for long storage is known as canning.

The term canning was first invented by N. Appert (1804) in France. In honour of the inventor canning is also known as appertization (Appart is known as the father of canning). The main principle of canning is destruction of spoilage organism by means of heat. The procedure of the canning should be as follows.

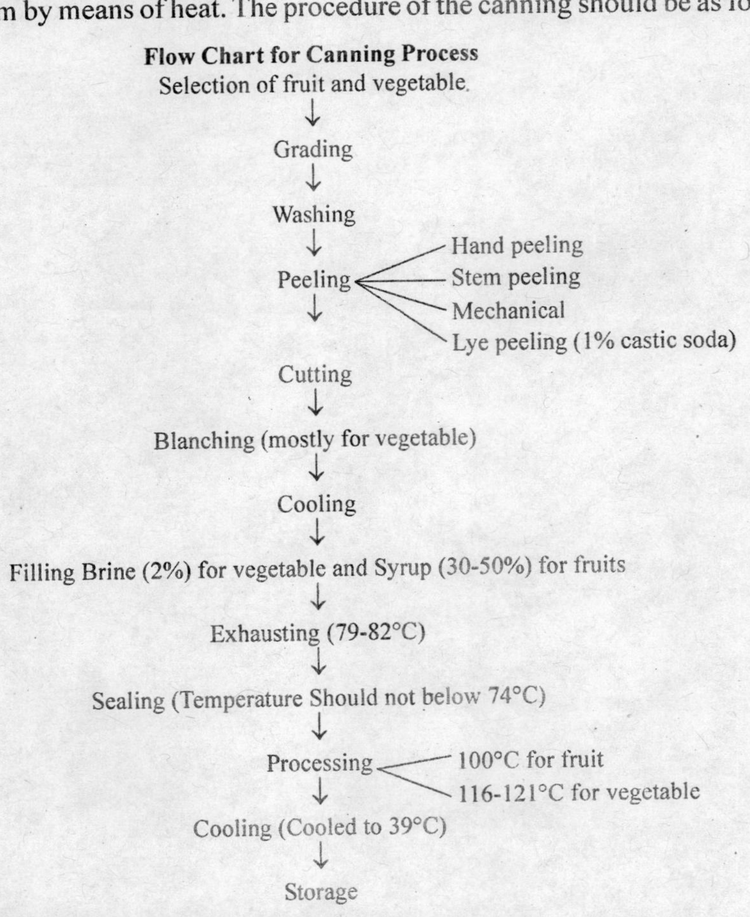

66

Jam

Jam:- Jam is a product made by boiling fruit pulp with sufficient sugar to a reasonably thick consistency. Jam is mainly prepared by apple, pear, sapota, apricot, peach, papaya, karonda etc. It can be prepared from one kind of fruits or from two or more kind of fruits. In general jam contain 0.5-0.6% acid and invert sugar should not be more than 40%.

Fruit and vegetable	Ingredient for 1kg pulp		
	Sugar (g)	Citric acid (g)	Water (ml)
Apple	750	2.0	100
Aonla	750	0	150
Apricot	600	1.0	100
Guava	750	2.5	150
Karonda	800	0	100
Mango	750	1.5	50
Musk melon	750	2.5	50
Pear	750	1.5	100
Peach	800	3.0	100
Plum	800	0	150
Mixed jam	800	2.5	100

Procedures of the jam preparation are as follows:

Flow Sheet for Processing of Jam

Selection of the fruit (Ripe firm fruits)

↓

Washing

↓

Peeling

↓

Pulping (Remove seed and core)

↓

Addition of sugar (add water if necessary)

↓

Boiling (with continous stirring)

↓

Addition of citric acid

↓

Judging end point — Temp. 105^0C / TSS 68-70 % / Sheet test

↓

Filling hot into sterilized bottles

↓

Cooling

↓

Waxing

↓

Capping

↓

Storage (at ambient temp.)

67

Jelly

Jelly:- Jelly is a semi solid product prepared by boiling a solution of pectin containing fruit extract free from pulp after the addition of sugar and acid. A perfect jelly should be transparent, well set, should have original flavour of the fruits, attractive colour and it should be free from dullness. Jelly is mainly prepared by guava, sour apple, wood apple, loquat and papaya. Apricot, pineapple, strawberry etc of low pectin containing fruits can also utilize after addition of pectin powder because these fruits have low pectin contain.

Fruit and vegetable	Ingredient for 1kg pulp	
	Sugar (g)	Citric acid (g)
Guava	750	3.0
Sour apple	750-1000	2.0
Wood apple	1000	0
Papaya	750	3.0
Jamun	750	1.0
Karonda	750	0
Plum	750	2.5
Loquat	800	2.0
Gooseberry	800	0

The procedures of the jelly preparation are as follows

Flow Sheet for Processing of Jelly

Selection of fruit (firm, not overripe)
↓
Washing
↓
Cutting into thin slices
↓
Boiling with water
(ratio of fruit: water is 1:1.5 for about 20-30min.)
↓
Addition of citric acid during boiling
(2g/kg of fruit)
↓
Straining of extract
(After keeping overnight)
↓
Pectin test — Alcohol test / Jelmeter test
↓
Addition of sugar
↓
Boiling
↓
Judging the end point — Temperature test / Sheet test / Drop/flake test
↓
Removal of scum or foam (1 teaspoon edible oil added)
↓
Colour and remaining citric acid added
↓
Filling hot in sterilized containers
↓
Waxing
↓
Capping
↓
Storage (at ambient temp.)

68

Marmalade

Marmalade: Marmalade is a fruit jelly in which peels of the fruits is in suspended forms. This product is generally prepared from citrus fruit like orange and lemon. The citrus marmalade can be classified into jelly marmalade and jam marmalade. 1 kg pectin extract required 750g sugar and 62g shredded peel.

The procedures of the marmalade preparation are as follows

Flow Chart for Processing of Marmalade

Selection of the fruits (Ripe fruit)
↓
Washing
↓
Peeling of the outer layer (Yellow portion)
↓
Cutting of the yellow portion into fine shreds
↓
Boiling (2-3 time in weight of the water for 40-60 minutes)
↓
Testing of pectin contain (alcohol test)
↓
Adding of sugar (as per requirement)
↓
Cooking (103-105^{0}C and continuous stirring)
↓
Addition of the prepared shred (shredded peels boiled for 10-15 minutes in several change of water for softening and removing bitterness and added @ 62g/kg of extract)
↓
Boiling till jellying point (continuous stirring)
↓
Testing for end point (Sheet/drop/temperature test)
↓
Cooling (82-88^{0}C with continuous stirring)
↓
Flavouring (orange oil)
↓
Filling
↓
Sealing
↓
Storage (at ambient temp.)

69

Preserve Fruit

Preserve:- A mature fruits/vegetable or its pieces impregnated with heavy sugar syrup till it become tender and transparent is known as preserve. The fruits like aonla, bael, apple, pear, mango, cherry, karonda, pineapple, papaya are commonly used.

Candied fruit:- A fruits/vegetable impregnated with cane sugar syrup and subsequently drained free of syrup and dried is known as candy. The most suitable fruits are aonla, apple, pear, cherry, karonda, pineapple, papaya are commonly used. In candy the total sugar contains of the impregnated fruits is kept about 75% to prevent fermentation.

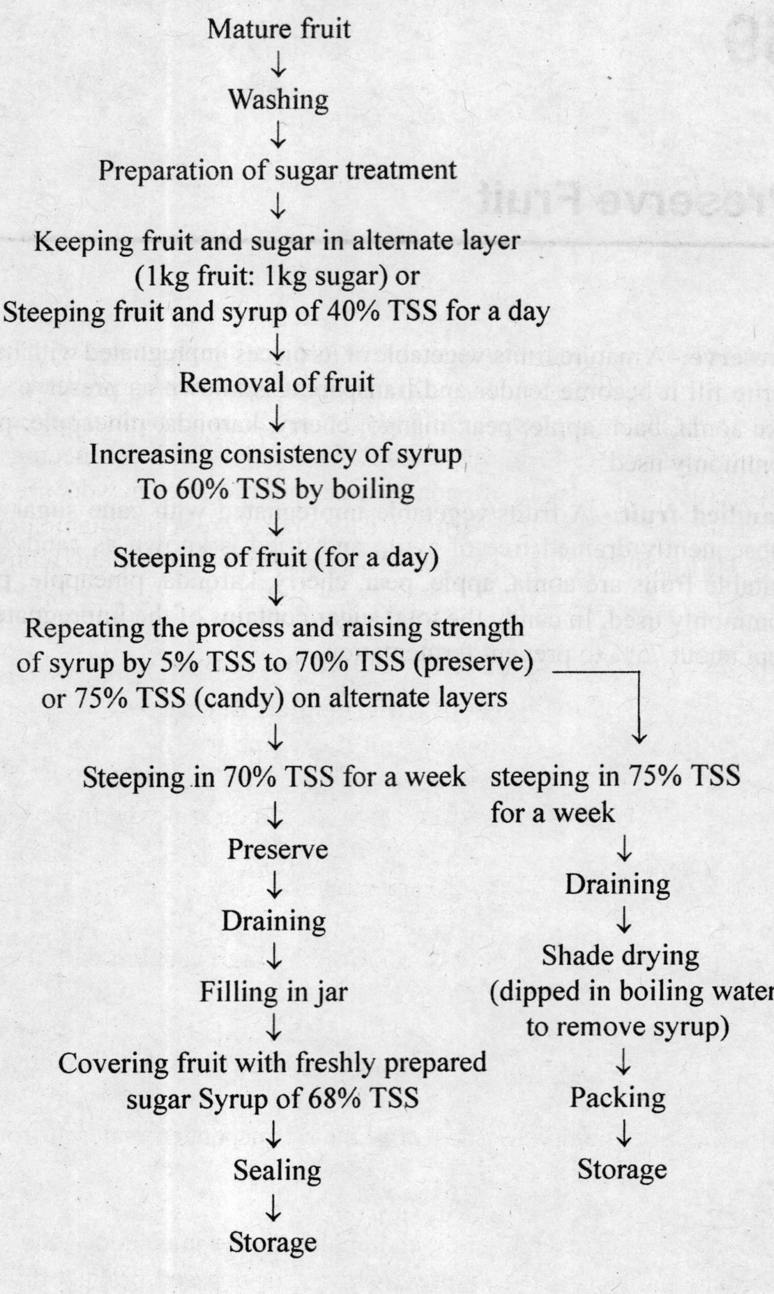
Flow Sheet for Preserve and Candy
Mature fruit
↓
Washing
↓
Preparation of sugar treatment
↓
Keeping fruit and sugar in alternate layer
(1kg fruit: 1kg sugar) or
Steeping fruit and syrup of 40% TSS for a day
↓
Removal of fruit
↓
Increasing consistency of syrup
To 60% TSS by boiling
↓
Steeping of fruit (for a day)
↓
Repeating the process and raising strength
of syrup by 5% TSS to 70% TSS (preserve)
or 75% TSS (candy) on alternate layers
↓
Steeping in 70% TSS for a week
↓
Preserve
↓
Draining
↓
Filling in jar
↓
Covering fruit with freshly prepared
sugar Syrup of 68% TSS
↓
Sealing
↓
Storage
steeping in 75% TSS
for a week
↓
Draining
↓
Shade drying
(dipped in boiling water
to remove syrup)
↓
Packing
↓
Storage

70

Fruit Beverage

Beverages:- The fruit beverages are easily digestible, highly refreshing, thirst quenching, appetizing, nutritionally for superior than many other synthetic drink. There are two types of fruit beverages i.e. unfermented beverages and fermented beverages. Natural juice, RTS (Ready to Serve), nector, cordial, squash, crush syrup, fruit juice concentrate and fruit juice powder are the main unfermented beverages. The fermented beverages are wine, champaigne, port, sherry, tokay, muscut, neera, cidar, orange wine, berry wine and fenny are the important.

Method of preparation of unfermented beverages

Flow Sheet of Unfermented Beverages

Selection of fruit (fully ripe fruit)

↓

Sorting and washing (diseased portion should be removed then washing)

↓

Juice extraction

↓

Deaeration (air should be removed by high vacuum with the help of dearater)

↓

Filtration (removal of the fruit skin, seed, broken tissue etc.)

↓

Clarification (complete removal of the all suspended material from the juice)

↓

Addition of sugar (except grape and apple) all the juice added the sugar for sweetening and sugar act as a preservative (sugar based product divided into three category i.e. low sugar [30% sugar or below], medium sugar [30-50% sugar], high sugar [50% or above sugar]) sugar can be added directly to the juice or as a syrup made by dissolving in hot water with small quantity of citric acid

↓

Fortification (addition of vitamins and ascorbic acid for enhancing the nutritive value)

↓

Preservation by pasteurization, cooling etc. and adding preservatives like KMS or Sodium Benzoate

↓

Bottling (bottle should be thoroughly washed with hot water and 1.5-2.5cm head space should be keep)

↓

Storage (at ambient temp.)

71

Squash

Squash: It is a type of fruit beverage containing at list 25% fruit juice or pulp and 40-50% TSS commercially. It's also containing about 1% citric acid and 350ppm sulphurdioxide or 600ppm sodium benzoate. Generally it is diluted before serving. Mango, orange, pineapple, lemon, lime, guava, litchi etc are commercially used for squash making using KMS as preservative. The coloured fruit like jamun, passion fruit, peach, phalasa, plum, mulberry, strawberry and grapefruits are also used for making squash but due to bleaching effect KMS is not used as preservative. In such types of fruits sodium benzoate used as preservative.

Fruit	Ingredient for one liter pulp/juice			
	Sugar (kg)	Water (litre)	Citric acid (g)	Preservative (g)
Lemon/lime	2.0	1.0	0	2.5 KMS
Orange	1.75	1.0	20	2.5 KMS
Litchi	1.8	1.0	25	2.25 KMS
Pineapple	1.75	1.0	20	1.9 KMS
Bael	1.8	1.0	25	2.5 KMS
Mango	1.75	1.0	20	2.5 KMS
Guava	1.8	1.0	20	2.0 KMS
Jamun	1.8	1.0	15	3.0 SB
Plum	1.9	1.0	10	4.0 SB
Phalasa	1.8	1.0	5	4.0 SB
Karonda	1.8	1.0	5	4.0 SB

Flow Sheet for Processing of Squash

Selection of fruits
↓
Washing
↓
Trimming
↓
Cutting or grating
↓
Juice extraction
↓
Straining
↓
Juice measuring
↓
Preparation of syrup
(Sugar + water + acid, heat just to dissolve)
↓
Straining
↓
Mixing with juice
↓
Addition of preservative
(0.6g KMS or 1.0g sod. benzoate/lit. of squash)
↓
Bottling
↓
Capping
↓
Storage

72

Guava Toffee

Material balance data

Guava pulp	: 5.00 kg
Amount of sugar added	: 4.50 kg
Amount of glucose added	: 500 gm
Amount of SMP added	: 700 gm
Amount of butter added	: 300 gm
Amount of Bengal gram flour added	: 714 gm

Flow Sheet for Processing of Guava Toffee

Fruits (Mature and fully ripe)

↓

Washing

↓

Peeling

↓

Cutting and grinding (removal of seeds)

↓

Pulping

↓

Heating

↓

Addition of sugar

↓

Addition of bengal gram flour, skim milk powder and butter

↓

Heating mixture till thick consistency

↓

Pouring mass in greased tray

↓

Drying at 70^{0}C for 4 hrs.

↓

Cutting into pieces of suitable size

↓

Wrapping in butter paper

↓

Filling in dry jar

↓

Storage

73

Pickle

Pickles:- Pickles are a product which is prepared by using the common salt or vinegars of fruits and vegetables. It is a good appetizer and adds to the palatability of a meal. It's also help full to stimulating the flow of gastric juices and thus helps full indigestion. Pickles are mainly preserving by salt, vinegar, oil and it can also be preserved by the mixture of salt, spices, vinegar and oil.

Ingredient of 1kg mango pickles: Mango piece 1kg, salt 150g, fenugreek powder 25g, turmeric powder 15g, nigella seed 15g, red chilli powder 10g, clove headless 8 number, black pepper 15g, cumin 15g, cardamom large 15g, aniseed powder 15g, asafoetida 2g and mustard oil 350ml.

Method of preparation of pickles

Flow Sheet for Processing Mango Pickle

Fruit (mature, green)
↓
Washing
↓
Cutting length wise into four pieces
↓
Removal of kernel
↓
Dipping pieces in 2% salt solution
(to prevent browning)
↓
Draining of water
↓
Drying in shade for few hours
↓
Heating oil
↓
Cooling
↓
Mixing spices in little oil
↓
Mixing with pieces
↓
Filling in jar
↓
Keeping in sun for a week
↓
Pressing the material (to remove air)
↓
Add remaining oil
↓
Storage

74

Mixed Pickle

Ingredient of 1kg mixed pickles: Cauliflower + diced carrot + turnip slice + vegetable pea each in equal amount of 1kg, salt 100g, ginger chopped 20g, onion chopped 50g, garlic chopped 10g, red chilli powder, black pepper, turmeric, cardamom large, aniseed powder, cumin, fenugreek powder each of 10g, clove headless 5 number and mustard 50g, vinegar 200 ml and mustard oil 450 ml.

Method of preparation of mixed pickle

Flow Sheet for Processing Mixed Pickle

Selection of vegetables

↓

Washing

↓

Trimming

↓

Peeling

↓

Cutting into slice or pieces

↓

Blanching

↓

Draining of water

↓

Drying in shed for an hour

↓

Frying spices in a little oil

↓

Contd.

Mixing piece with fried spices

↓

Cooling

↓

Making taste of jaggery and vinegar

↓

Adding paste to pieces

↓

Filling in jar

↓

Keeping in sun for a week

↓

Addition of oil after heating and cooling

↓

Storage

75

Tomato Sauce/Ketchup

Sauce/Ketchup:- Fully ripe red fruits are selected, all green and blemish part are discarded. It is made from strained tomato juice or pulp and spices, salt, sugar and vinegar with or without onion and garlic and contains not less than 12% tomato solid and 25% total solid.

Ingredient of tomato sauce/ketchup: Tomato pulp 1kg, sugar 75g, salt 10g, onion chopped 50g, ginger chopped 10g, garlic chopped 5g, red chilli powder 5g, cinnamon, cardamom large, aniseed, cumin, black pepper powder 10g each, clove headless 5 number, vinegar 25ml or glacial acetic acid 5ml and sodium benzoate 0.25g/kg final product.

Flow Sheet for Processing Tomato Sauce/Ketchup

Selection of tomato fruits (fully ripe and red)

↓

Washing

↓

Sorting and trimming

↓

Cutting and chopping

↓

Heating at 70-90^{0}C for 3-5 minutes (for softening)

↓

Pulping or extraction of juice/pulp (by mechanically or by sieving)

↓

Straining tomato pulp/juice

↓

Cooking pulp with 1/3 quantity of sugar

↓

Putting spice bag in pulp and pressing ocassionally

↓

Cooking to 1/3 of original volume of pulp/juice

↓

Removal of the spice bag (after squeezing in pulp)

↓

Addition of the remaining sugar and salt

↓

Cooking

↓

Judging of end point (tomato solid by hand refrectometer/volume by measuring stick i.e. 1/3 of its original volume)

↓

Addition of vinegar/acidic acid and preservative

↓

Feeling into bottles at about 88^{0}C

↓

Sealing (crown corking)

↓

Pasteurisation (at 85-90^{0}C for 30 minutes)

↓

Cooling

↓

Storage (at ambient temp.)

76

Tomato Chutney

Ingredient of tomato chutney: Tomato 1kg, sugar 500g, salt 25g, onion chopped 100g, ginger chopped 10g, garlic chopped 5g, red chilli powder 10g, cinnamon, cardamom large, aniseed, cumin, black pepper powder 10g each, vinegar 100ml and sodium benzoate 0.50g/kg final product.

Flow Sheet for Processing Tomato Chutney

Selection of tomato fruits (fully ripe and red)

↓

Washing

↓

Sorting

↓

Blanching for 2 minutes

↓

Putting in cold water (to crack skin)

↓

Peeling

↓

Crushing

↓

Addition of ingradient except salt and vinegar and cooking gently to desired consistancy

↓

Addition of salt and vinegar and cooking for 5 minutes

↓

Addition of preservative

↓

Filling hot into bottle

↓

Sealing

↓

Storage (at ambient temperature in cool and dry places)

77

Turmeric Processing

Structure of turmeric process

Boiling

- Boiling figures in water in copper vessels/galvanized iron/earthen vessels
- If water is acidic then add $NaCO_3$
- Stopped boiling when froth starts coming out and white fume appears
- Then dried in the sun for 10-15 days till they become hard and brittle

Polishing

- They polished mechanically in a drum rotated by hand
- On large scale they polished in a hexagonal wooden drum mounted on central axis and rotated by power

Colouring

- Colouring is done for giving a attractive colour
- For colouring figures are taken in a basket which is shaken continuously while a prepared emulsion is poured in
- Figures are uniformly coated with the emulsion
- They may be dried in the sun

Composition of the emulsion

Alum - 0.04 kg

Turmeric powder - 2.00 kg

Castor seed- 0.14 kg

Sodium bisulphate- 30 g

Concentrate HCl- 30 ml

This composition of the emulsion as recommended by the CFTRI, Mysore for 100 kg of polished turmeric.

78

Curing of Ginger

Curing- Curing is a process in which produce is subjected to undergow through a special treatment to get the quality product.

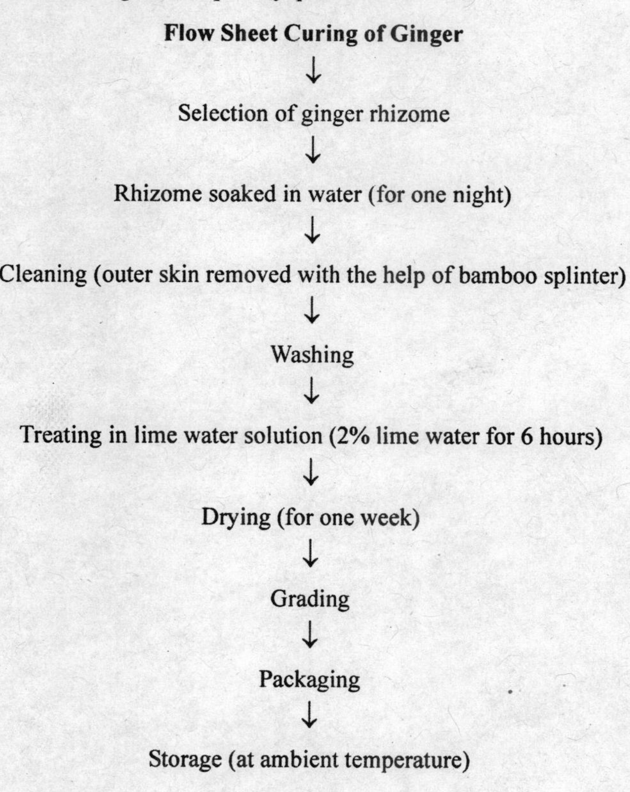

79

Distillation of Rose Oil and Water

Distillation of rose oil and rose water:

For distillation of rose oil and rose water initially the fresh rose flower are hydro distilled. The oil and water are collected together. The oil which floats upon the water is separated by decantation get oil directly, while the water is re-distillated. The oil obtained from re-distillation is called recovered oil or water oil. The direct oil and water oil are then combined to make the commercial rose oil.

The water obtained from the second re-distillation step is the commercial rose water

Flow Chart for Distillation of Rose

Fresh rose flower 350kg

↓ Distillation with water (7 hours)

Residue discharge | Distilled water (First Rose water 350 Lit) | Direct/Decanted oil (15g oil)

↓ (from Distilled water)

Four batches of first rose water stored1400 Lit

↓

Re-distillation in one batch (for 6 hour)

↓

Residual water (spent rose water 1000 lit) | Rose water distilled (Second rose water) | Recovered oil or water oil (75g oil)

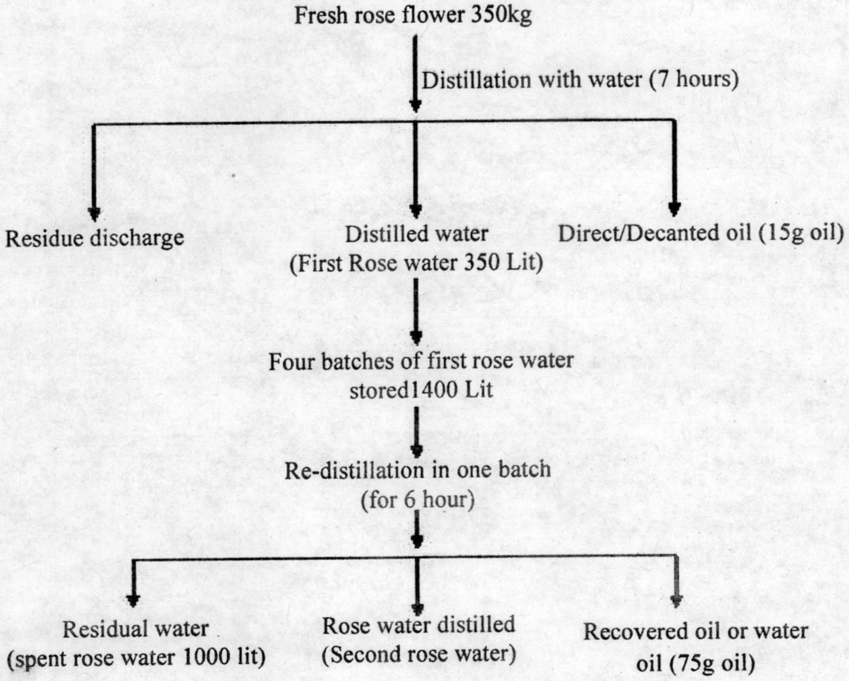

79

Distillation of Rose Oil and Water

Distillation of rose oil and rose water :

For distillation of rose oil and rose water initially the fresh rose flower are hydro distilled. The oil and water are collected together. The oil which floats upon the water is separated by decantation get oil directly, while the water is re-distillated. The oil obtained from re-distillation is called recovered oil or water oil. The direct oil and water oil are then combined to make the commercial rose oil.

The water obtained from the second re-distillation step is the commercial rose water.

Flow Chart for Distillation of Rose

Formate for Thesis Writing, Research Paper, Review Article etc.

The practical exercise is written under the following heads: Give the done on which the practical exercise was started on the left hand corner (margin) of the record book

1. **Title:** Give a brief suitable title to each practical exercise which should indicate the contents of the experiment.
2. **Introduction:** It should deal, in brief the nature, purpose, scope and Justification of the experiment.
3. **Review of the Literature:** It should deal, in brief, summary of the works reported on the problem or problems of similar nature. State the technique used, the year of publication (in bracket) and the author's name. Include sufficient number of reviews to act as guideline in planning the experiment. Underline the generic and species names.
4. **Methods and Materials:** Describe the equipments and materials used with their specifications. Give the design and layout used along with other details of the experiment in detail. The method followed in the investigation should be described in detail.
5. **Experimental Findings:** Experimental data are recorded in this section. The findings of the experiment may be presented with the help of tables, photograph, graphs and captioned diagrams. A suitable title stating the contents clearly and concisely should be given for each table. Analyse the data adopting a suitable statistical method.
6. **Discussion:** A brief discussion of the results of the investigation should be given on the basis of findings of other workers or the similar problem as already cited in the 'review of literature'. Present the evidence of each finding Try to account for unexpected results and exceptions.

7. **Summary and Conclusion:** It should deal with the findings of the experiment in brief without missing any item of the finding and design of experiment.

8. **References:** All reference cited in the 'review of literature' or elsewhere be listed One such procedure for listing reference is given below.

1. Book

Author's Name, year of publcation (in bracket) title, of the book, edition, publishers name with address, page (s) used.

Example

Bal, J.S. (2006). Fruit Growing, 2nd Ed., Kalyani Publishers, Ludhiana, pp. 100-125 or P. 125.

2. Journal

Author's name, year of publication (in bracket), title of the article, journal's name, volume, number (in bracket), page (s).

Example

Singh, B. K., Pal, A. K., Singh, A. K. and Verma, A. (2016). Impact of integrated nutrient management on chlorophyll and nutrients concentration in strawberry leaf cv. Chandler. *Green Farming*, 7 (4): 1-3.

3. Bulletin

Author's name, year of publication (in bracket). Bulletin's name and place of publication, number, page (s).

Example

Pople, W. T. (1929) "Mango Culture in Hawaii", Agri, Expt. Stn, Bull. No. 58, pp. 60-62

4. Newspaper

Anonymous, year (in bracket), title of the article, place of publication, name of the paper, date, page (s) and column.

Example

Anonymous (1959). Guava Wilt, Allahabad. "The Leader" Feb. 10, 1965, p.5, Col. 6.

5. Chapter of a Book

Authors(s) name year of publication (in bracket), title of the chapter, page (s), name of the editing author (s), title of the book, editor's name and place of publication.

Example

Fox, J.E. (1969). The cytokinin, pp. 85-123 in M.B. Wilkin, Ed. The physiology of plant growth and development, McGraw- Hill Publishing Company, New York.

6. Abstract

Author (s) name (s), year of publication, title of the article, abstract number, name of the abstract, name and place of publication.

Example

Singh, B.K., Verma, A. and Singh, R.K. (2013). Role of Drip Irrigation in Grape (*Vitis vinifera*), National Seminar on Enhancing Water Productivity in Agriculture, C.A.B. No. **177,** Horticultural Abstract, B.H.U., Varanasi.